Capsicum - Current Trends and Perspectives

Edited by Orlex Baylen Yllano

Published in London, United Kingdom

Capsicum - Current Trends and Perspectives
http://dx.doi.org/10.5772/intechopen.98137
Edited by Orlex Baylen Yllano

Contributors
Oluyinka Adewoyin, Adebayo Ibidapo, Amos Famaye, Rufus Ipinmoroti, Folasayo Fayose, Felipe R. Garcés Fiallos, Anthony A. Moreira Morrillo, Álvaro Monteros-Altamirano, Ailton Reis, Dalasanuru Chandregowda Manjunathagowda, Raman Selvakumar, Praveen Kumar Singh, Mirunalini Sankaran, Kalaiyarasi Dhamodharan, Manobharathi Vengaimaran, José Díaz, Raquel Núñez-Fernández, Javier Veloso, Aman Dekebo, Orlex Baylen Yllano, Nkwiza M. Nankolongo, Leilani D. Dela Roca Arce, Neil John V. Vegafria, Epharaim A. Evangelista, Ferdinand A. Esplana, Lester Harris R. Catolico, Merbeth Christine L. Pedro, Edgar E. Tubilag

First published in London, United Kingdom, 2023 by IntechOpen
IntechOpen is the global imprint of INTECHOPEN LIMITED, registered in England and Wales, registration number: 11086078, 5 Princes Gate Court, London, SW7 2QJ, United Kingdom

British Library Cataloguing-in-Publication Data
A catalogue record for this book is available from the British Library

Additional hard and PDF copies can be obtained from orders@intechopen.com

Capsicum - Current Trends and Perspectives
Edited by Orlex Baylen Yllano
p. cm.
Print ISBN 978-1-80356-005-2
Online ISBN 978-1-80356-006-9
eBook (PDF) ISBN 978-1-80356-007-6

Meet the editor

Dr. Orlex Baylen Yllano is a professor and chair of the Department of Biology, College of Science and Technology, Adventist University of the Philippines. He holds a Ph.D. in Molecular Biology and Biotechnology and completed post-graduate research in Japan, Thailand, and the United States. Dr. Yllano has published papers on crop molecular biology and biotechnology, phytoremediation, and conservation genetics. He has mentored graduate and undergraduate thesis students. He has published research papers in Scopus/ISI-indexed journals and co-authored several books. He is affiliated with national and international scientific societies. He serves as an editorial board member of an online journal and a reviewer for several other journals. He is a recipient of numerous research and academic awards. Above all, Dr. Yllano has a passion for research, teaching, and mentoring.

Contents

Preface

Over the centuries, *Capsicum*s have been cultivated and transformed into various shapes, sizes, and flavors and have become essential ingredients in diverse cuisines, supplements, and functional foods. The economic potential of *Capsicum* makes it a valuable crop in many parts of the world.

Chapter 1, "Genetics and Genomics of Capsicum: Valuable Resources for Capsicum Development" by Nkwiza M. Nankolongo, Orlex Baylen Yllano, Leilani D. Arce, John Neil V. Vegafria, Ephraim A. Evangelista, Ferdinand A. Esplana, Harris Lester R. Catolico, Merbeth Christine L. Pedro, and Edgar E. Tubilag, provides valuable perspectives on *Capsicum's* biology and horticultural characteristics, genetic resources, genetic diversity, phylogenetic relationships, ploidy levels, chromosome structures, genome organization, important genes, and their applications. The study of *Capsicum* genetics and genomics is critical for this crop's continued improvement and sustainability.

In recent years, Padron peppers have attracted the attention of *Capsicum* enthusiasts worldwide due to their unique texture and flavor. Chapter 2, "Padrón Peppers, Some Are Hot, Some Are Not" by José Díaz, Raquel Núñez-Fernández, and Javier Veloso, highlights capsaicinoids' expression, biosynthesis, metabolic pathways, and homeostasis. These pathways are modulated by different factors, namely, plant hormones, transcription factors, ontogeny, and the environment, including both abiotic and biotic agents.

The discovery and elucidation of bioactive compounds in peppers have paved the way for advancing new therapeutics and functional food. Chapter 3, "Pharmacological Properties and Health Benefits of *Capsicum* Species: A Comprehensive Review" by Kalaiyarasi Dhamodharan, Manobharathi Vengaimaran, and Mirunalini Sankaran, emphasizes *Capsicum's* essential constituents and bioactive compounds. This chapter highlights the pharmacological uses and potential health benefits of *Capsicum* species and their active compounds in diverse aspects.

Like any other plant, *Capsicum* is the target of pathogens like viruses, bacteria, and fungi. Among the destructive pathogens of *Capsicum* is *Phytophthora capsici*, which can cause significant damage to the crop and reduce productivity. Chapter 4, "*Phytophthora capsici* on *Capsicum* Plants: A Destructive Pathogen in Chili and Pepper Crops" by Anthony A. Moreira-Morrillo, Álvaro Monteros-Altamirano, Ailton Reis, and Felipe R. Garcés-Fiallos, addresses the etiology, symptomatology, occurrence, cycle, and management of disease caused by *P. capsici*.

Pest management in *Capsicum* production is crucial to maintaining high crop yields and quality. Chapter 5, "Major Pests and Updates on Pest Management Strategies in the Sweet Pepper (*Capsicum annuum*)" by Aman Dekebo Aman, dissects the sustainable and alternative pest management strategies to control pests of sweet pepper. These strategies include intercropping sweet pepper with plants against pests, oviposition deterrents, natural enemy release, use of resistant cultivars, and eliciting plant defenses as environmentally friendly control methods.

With the proliferation of pests that target the different *Capsicum* species and the impact of climate change, efficient and sustainable breeding strategies must be a priority. Chapter 6, "Capsicum: A Breeding Prospects and Perspectives for Higher Productivity" by Dalasanuru Chandregowda Manjunathagowda, Raman Selvakumar, and Praveen Kumar Singh, examines the *Capsicum* breeding programs, which aim to improve yield, biotic, abiotic resistance, and nutritional quality. The chapter elucidates the recent breakthroughs in *Capsicum* breeding and the role of classical selection and hybridization procedures coupled with molecular and genetic technology in establishing a more robust *Capsicum* breeding program.

Postharvest management of peppers is a critical step in ensuring the quality of the produce. Peppers are highly perishable and susceptible to damage during harvesting, handling, and storage, resulting in significant yield, quality, and market losses. To minimize huge postharvest losses, there must be efficient postharvest handling, processing, and preservation methods compatible with the socioeconomic and cultural practices of the producers. Chapter 7, "Postharvest Handling Methods, Processes, and Practices for Pepper" by Oluyinka Adewoyin, discusses postharvest methods to enhance the adequate supply of pepper fruits and stabilize their price for the benefit of both producers and consumers.

The information discussed in this book is indispensable in *Capsicum* frontier research, breeding, development, management, and utilization of this economically important and highly regarded crop worldwide. Join us on a journey through the latest trends and perspectives in *Capsicum* and discover what this history-rich and valuable crop has to offer.

I would like to relay my sincere gratitude to all the authors for sharing their valuable works. I hope this book will inspire readers to explore *Capsicum* as a promising crop that offers many potential benefits.

[Medical Disclaimer: The information in this book should not be used in place of expert medical advice, diagnosis, and treatment.]

Dr. Orlex Baylen Yllano
Professor and Chair,
Department of Biology,
College of Science and Technology,
Adventist University of the Philippines,
Puting Kahoy, Silang, Cavite, Philippines

Chapter 1

Genetics and Genomics of Capsicum: Valuable Resources for Capsicum Development

Nkwiza M. Nankolongo, Orlex Baylen Yllano, Leilani D. Arce, Neil John V. Vegafria, Ephraim A. Evangelista, Ferdinand A. Esplana, Lester Harris R. Catolico, Merbeth Christine L. Pedro and Edgar E. Tubilag

Abstract

Capsicum is a genetically diverse eudicot, diploid, and self-pollinating plant that grows well in slightly warmer environments. This crop is popular in different areas of the world due to its medicinal properties and economic potential. This chapter evaluated and analyzed the *Capsicum's* biology and horticultural characteristics, genetic resources, genetic diversity, phylogenetic relationships, ploidy levels, chromosome structures, genome organization, important genes, and their applications. This chapter is indispensable in *Capsicum* frontier research, breeding, development, management, and utilization of this economically important and highly regarded crop worldwide.

Keywords: *Capsicum*, genetic diversity, genome, chromosomes, ploidy levels

1. Introduction

Capsicum is an economic crop cultivated worldwide for spice in a wide array of cuisines, ornamental plants, source of vitamins, minerals, bioactive compounds, biopesticides, components of cosmetics, and other indigenous, medicinal, and industrial uses. The popularity of *Capsicum* caught the attention of breeders, researchers, and enthusiasts to propagate and develop this wonder crop. The number of *Capsicum* cultivars and taxonomic varieties is increasing [1, 2]; however, the genus *Capsicum* has five domesticated species and around 25 identified species [3]. These five *Capsicum* taxa (*C. annum, C. baccatum*, *C. chinense*, *C frutescens,* and *C. pubescens*) can be differentiated through their morphological characteristics like bloom and seed color, calyx form, number of flowers per node, and flower orientation [1, 3, 4]. Interestingly, *C. annum* is considered to have been domesticated from *C. annum* populations in the wild, while *Annum glabriusculum* in Mexico was derived from many geographically distinct wild populations [3].

2. *Capsicum* biology and horticultural characteristics

The CABI Compendium features *Capsicum's* biology and horticultural characteristics as follows [5]. *Capsicum* grows to a height of 0.5–1.5 m, is heavily branched, and has extremely strong taproots. The lateral roots are many, while the stem is uneven and angular, measuring around 1 cm in diameter and measuring about 0.5–1.5 m in length. The stem is normally green to brown-green in color, with purple patches near the node on occasion. The leaf design is alternate, basic, and highly changeable, with petioles up to 10 cm long. The apex is acuminate, and the edge is whole, pale dark green, and subglabrous. The flowers are arranged singly, and the pedicel is around 4 cm long when in bloom. The fruits may grow up to 8 cm in length. The calyx is cup-shaped and with enlarged fruit. Generally, it has five conspicuous teeth, and the white corolla has five to seven lobes. It can develop to five to seven stamens with pale blue to purplish anthers. The ovary is 2–4 locular, style filiform, and has a white or purplish stigma capitate. Fruit is a non-pulpy berry variable in size, shape, color, and degree of pungency [5]. Fruits are relatively conical with up to 30 cm long. The colors tend to be green, yellow, cream, or purplish when it is not yet fully developed. However, it becomes red, orange, yellow, or brown when it matures. The seeds are orbicular and flattened, about 3–4.5 mm in diameter, 1 mm thick, and pale yellow in color. The plant is considered an annual, herbaceous, perennial, seed propagated, shrub, and climber.

Domesticated crop seeds germinate 6–21 days after seeding, with continuous blooming beginning 60–90 days later. The flower is open for 2–3 days, and outcrossing of up to 91% may occur, depending on bee activity and heterostyly, although it is typically considered a self-pollinated crop. Approximately 40–50% of the flowering set fruit matures 4–5 weeks after blooming and can be plucked in 5–7 days intervals under typical conditions. The harvest time is around 4–7 months following the seeding stage [5].

The untamed seed dormancy in wild *C. annum* seeds is staggered, allowing germination and recruitment to occur when conditions are best in a more unpredictable and uncertain environment [6]. Wild seeds' testae are thicker than domesticated plants, generating more but smaller seeds, which are better equipped for dispersion [6]. Insect pollinators outcrossed *C. annum* in the wild at a significant rate were noted [6]. Flowering occurs late in the season, but once it does, it is persistent and prolific, with overlapping stages of flower and fruit development [6]. *Capsicum* plants may grow as permanent shrubs in a suitable climate, although they are commonly grown as annuals elsewhere [6]. Light, well-manured, limey, and well-drained soil is preferred [1].

Capsicum peppers are day-neutral, warm-season plants; however, certain varieties may exhibit a photoperiodic reactivity [5]. The vegetative cycle may be accelerated by enforcing particular photoperiods [5]. *Capsicum* peppers can withstand 45% of prevailing sun energy in a shaded environment [5], although shadow might delay flowering.

Capsicum peppers thrive in loam soil with a pH of 5.5–6.8 [5]. They grow at various elevations, with rainfall ranging from 600 to 1250 mm [5]. Cultivars are destroyed by severe floods or drought [5]. The optimum germination temperatures are between 25 and 30°C and can withstand temperatures as low as 15°C at night [5]. The plant can yield fruit, albeit it will be delayed if the temperature drops below 25°C [5]. If the night temperature hits 30°C, flower buds will abort rather than mature [5]. When the temperature reaches 30°C, and below 15°C, the pollen viability is significantly reduced [5].

Capsicum seeds are dispersed in various ways. *C. annum* is transmitted via the movement of seeds, which are generated in vast quantities and can endure for more than a year [3]. Chilies are the favorite food of many birds in their natural range; they drop seeds while eating the fruits or pass through the digestive tract unharmed [3]. Humans also intentionally spread the species to use its fruits and leaves as food, spice, ornamental, and medicine [1, 7]. It is believed to have escaped cultivation accidentally in Puerto Rico and Finland [8, 9]. Because the species can grow in sandy, coastal environments, it can be spread by both biotic and abiotic vectors [10].

3. Ploidy levels and chromosome structure

Studies of ploidy levels and chromosome structure of *Capsicum* provide essential data for *Capsicum* taxonomy, assisting in the identification of cultivated, semi-cultivated, and wild species, as well as contributing to plant variety improvement and conservation [11].

Capsicum species are diploids, and most of them have 24 chromosomes (n = x = 12), but several wild species have 26 chromosomes (n = x = 13) [11, 12]. *C. annum* has 24 chromosomes; usually, two pairs are acrocentric, and 10 or 11 pairs are metacentric or submetacentric [13]. Its nuclear DNA content has 3.38 picograms (pg) per nucleus, which, in relation to other reports, ranged from 2.76 to 5.07 pg. per nucleus [14]. The chili pepper genome ranged from 1498 cm to 2268 cm and approximately two to three times larger than the tomato genome [15, 16].

In 12 *Capsicum* accessions, a chromosome number of 2n = 2x = 24 was determined, and this ploidy level is well-documented in several *Capsicum* species [17]. *Capsicum* species in the wild, such as *C. buforum,* has a ploidy level of 2n = 2x = 26 [17]. Two distinct evolutionary lines emerged throughout the history of this genus, marked by a significant separation between wild (base number x = 13) and domesticated (base number x = 12) species [17]. Multiple karyotypic formulae in the same species may occur from genetic variances within populations, which are produced by genomic responses to various environments [17]. Individuals in the same group might have different chromosomal races due to chromosomal polymorphism [17].

Plants with more karyotypic symmetry than other members of the same genus are related to those with less symmetry [18]. Even though most *Capsicum* species have 2n = 24 and have quite similar chromosomal shapes, the genus exhibits a lot of intraspecific and interspecific karyotypic diversity [18].

Karyotypic asymmetry is linked with considerable changes in TLHB and TCL across individuals of the same or nearly related species due to chromosomal modifications such as Robertsonian translocation, inversion, uneven translocations, deletions, and duplications [18]. Exposure to external elements such as climate, soil, temperature, and moisture may cause these changes [18]. *C. annum* and *C. chinense* chromosome modifications like translocations, duplications, and deletions have been identified [18]. The karyotypes of *Capsicum* in the same genus showed more genetic variability, possibly due to their high asymmetry index [19].

Studies of pepper chromosome number and morphology produce essential data for *Capsicum* taxonomy, which aid in delineating cultivated, semi-cultivated, and wild species and contribute to plant diversity conservation by providing valuable information for breeding and genetic improvement programs of this crop [20, 21].

The diploid chromosomes (2n = 2x = 24) were confirmed for each of the 12 accessions. This ploidy level is well-documented in several *Capsicum* species [19, 22–26]. For some wild *Capsicum* species, such as *C. buforum* and *C. capylopodiume,* 2n = 2x = 26 ploidy level has been reported [24]. Throughout the development of this genus, two separate lines arose, as evidenced by a clear divergence between wild (base number x = 13) and domesticated (base number x = 12) species [24]. It was hypothesized that x = 13 lines are inherited from ancestors of the x-12 plants [24].

The karyotypic formula 11M + 1SM was determined in 11 of the examined accessions, with chromosome 12 categorized as submetacentric [19]. The karyotypic formula 12M was observed in the *C. frutescens* accessions BGC 37, indicating chromosomal polymorphism compared to the other accessions [19]. For several Venezuelan accessions, the karyotypic formula 11M + 1A was reported [22]. At the same time, the formula 11M + 1A in *C. chinense* accessions was described through conventional cytogenetics in several Brazilian states [20, 27].

Genetic differences within populations, caused by the genomic response to diverse environments, might result in multiple karyotypic formulae in the same species [25]. Chromosomal polymorphism might change the karyotypic pattern of individuals in the same group, resulting in separate chromosomal races [19].

Variances in the form, size, and number of chromosomes are prevalent in populations of the same species or interspecific taxa. These differences are categorized into cytotypes or chromosomal races [19, 20]. Researchers confirmed that such variations are common in the *Capsicum* genus, whose cytotypes differ primarily in karyotypic formula and chromosomal size [19]. Secondary constrictions were found in the homologous pairs (1 and 12; 6 and 11) of the BGC 01 and BGC 37 *C. frutescens* accessions, respectively [19]. Prominent secondary constrictions were found in every *Capsicum* species, ranging from one to four per karyotype [27]. The average chromosomal size measured in various *Capsicum* species ranged from 3.29 m (BGC 49) to 7.48 m (BGC 54) [19].

Most of the *Capsicum* species have similarity (2n = 24), and the genus also exhibited intra- and interspecific karyotypic variability [28]. A higher asymmetry index across karyotypes of species in the same genus is associated with more genetic heterogeneity [20]. *Capsicum*'s chromosomal analysis at the metaphase stage revealed metacentric, submetacentric, acrocentric, or telocentric chromosomes [22, 29].

Capsicum species had symmetrical chromosomal numbers, so more extensive sampling and detailed characterization of the chromosomes, including heterochromatin distribution and sequence identification by *in situ* hybridization, must be done to distinguish between species that have the same karyotypic formula [19, 30].

Studies on pepper chromosome number and shape provide essential data for *Capsicum* taxonomy, assisting in the identification of cultivated, semi-cultivated, and wild species, as well as contributing to plant variety conservation by assisting genetic improvement efforts for this genus [31].

4. Genetic diversity and QTLs

Genetic diversity is crucial for *Capsicum* development and management. Genetic diversity is correlated with average fitness across populations [18, 32]. The robust genetic diversity of *Capsicum* populations will enable them to adapt to ever-changing environments, be a resource of valuable alleles and genes in the population, and contribute to ecosystem diversity [18, 32]. To date, there are various estimates as to the number of *Capsicum* species [20–24, 33–35].

Capsicum fruits are diverse and vary in form and characteristics. Fruit form has been examined extensively in the Solanaceae family, including tomato, pepper, and eggplant [36–38]. On chromosomes 2, 3, 7, 8, and 10, allelic variations in the Sun, Ovate, Fascinated (FAS), and Locule Number (LC) genes determine the form of the tomato fruit [36, 39–46]. Individual alleles of these genes might account for up to 71% of the particular shape variance in a population [46]. Individual alleles of these genes accounted for up to 71% of the particular shape variation in a broad sample of 368 wild and cultivated tomatoes [46]. Fruit weight was strongly co-localized across tomato and pepper QTLs, and a single fruit shape QTL was co-localized, suggesting that conserved components contribute to one, if not both, of the traits [38, 39, 44–51]. Multiple QTLs for fruit length, width, and the fruit shape ratio (length: width) have been discovered on chromosomes 1–4, 8, 10, and 11 [39, 44, 48, 50]. Two essential fruit QTLs, fs 3.1 (fruit shape) and fe 10.1 (fruit elongation), were linked to chromosome 3 and 10, respectively, in a BC4F2 population segregating for fruit-shaped [44]. These QTLs accounted for 67, and 44% of the variance in fruit form and elongation found in the population, respectively [44]. Fruit trait inheritance in connection to pericarp form, color thickness, and total soluble solids was also investigated [52].

The round form characteristic was governed by a single gene based on segregation ratios. Five QTLs contributing to fruit form and one QTL for pericarp thickness on chromosomes 1, 2, 4, 10, and 3 were determined. This explains the 4 to 26% of the diversity in the Jalapeno recombinant inbred lines [50]. The expression of a gene that resembled the tomato gene Ovate and discovered substantial variations between round and elongated pepper cultivars was compared [51]. Further study was carried out on five domesticated species [53].

Another QTL analysis in 2012 found that two dominant genes regulated fruit mass length, diameter, form ratio, and flesh thickness, with heritability ranging from 38 to 88% [45, 54]. Fruit width was highly heritable, and fruit weight and width were positively associated while examining a pepper germplasm collection from the Caribbean, which was consistent with the QTL study [44, 48, 54]. The heritability of fruit form and flesh thickness was 80% in another mapping investigation [48]. The INRA characterized the phenotype of almost 1300 pepper accessions in their collection for 12 fruit traits; form and color varied across domesticated species, but wild species often featured tiny, elongated fruit [55]. Despite the large number of studies examining pepper fruit form, the use of subjective visual (e.g., elongate, triangular, square, heart) or manual (length/width ratio) measures to define fruit shape was a shortcoming in all of them [55]. With this, software has been created that allows for more objective and reliable assessments of fruit attributes [33, 41].

Disease resistance is also crucial, even if fruit form is one of the most critical features of a cultivar [30]. Cultivated cultivars frequently lack disease resistance due to breeding bottlenecks [30]. Resistance is frequently found in small-fruited wild species and then adopted into larger-fruited commercial cultivars [56, 57]. Through linkage drag or pleiotropic effects, negative horticultural features such as disease resistance can be passed together with beneficial traits like disease resistance [56, 57]. Recent research on tomatoes found a relationship between resistance to the late blight pathogen (*Phytophthora infestans*) and unfavorable impacts on maturity, fruit size, yield, and plant architecture [58, 59]. In pepper, a link between fruit features and disease resistance for a single strain of *P. capsici*, a destructive fungus that causes fruit, foliar, and root rot [50, 55]. In an eggplant germplasm population, fruit form was positively linked with disease susceptibility to *P. capsica* [48, 50]. Negative associations were found between resistance to the bacterium pathogen *Pseudomonas syringae* (PV) in Kiwi [60]. When transferring disease resistance into commercial cultivars, it is critical to look for possible correlations, linkage drag, and pleiotropic effects [60].

Since their introduction in Mexico, peppers have been subjected to a substantial selection for fruit forms and sizes [61, 62]. Domesticated pepper fruit has an unlimited variety of phenotypic variability [63–66]. While cousins and landrace peppers are typically tiny and very pungent, domesticated pepper fruit has an endless array of phenotypic diversity [64–66]. Various regional preferences exist for pepper consumed in most nations and marketplaces [64]. Regional choices have increased morphological variability among market classes [64].

In addition to Capsicum's genetic diversity, we also analyzed the phylogenetic relationships of 22 Nicotinamide Adenine Dinucleotide Dehydrogenase (NADH dehydrogenase) sequences of Capsicum from 19 different species. NADH dehydrogenase is a flavoprotein-containing oxidoreductase that catalyzes the conversion of NADH to NAD. The enzyme may be found in eukaryotes as a part of the mitochondrial electron transport complex I and in transferring electrons from photoproduced stromal reductants like NADPH and ferredoxin to the intersystem plastoquinone pool. NADH dehydrogenase is the primary enzyme complex in the electron transport chain in mitochondria. Nicotinamide adenine dinucleotide (NAD) is transformed from its reduced form, NADH, to its oxidized form, NAD+ [67].

Phylogenetic analysis using UPGMA Maximum Composite Likelihood with 1000 bootstrap replications revealed that the *C. ciliatum, C. lanceolatum, C. lycianthoides,* and *C. geminifolium* grouped together. More so, *C. minutiflorum and C. ceratocalyx* were strongly clustered together, the same with *C. chinensis and C. frutescens* (**Figure 1**).

C. pubescens, C. galapagoence, C. chacoense, and *C. cardenasi* formed a cluster adjacent to *C. annuum* on top and *C. baccatum* species below (**Figure 1**). Other *Capsicum*

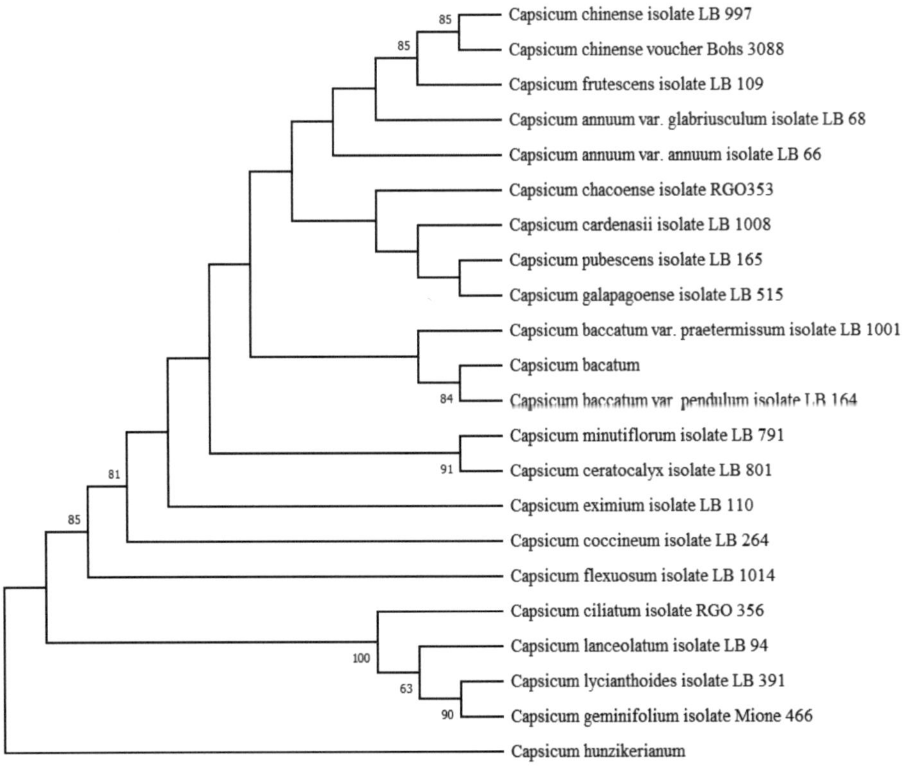

Figure 1.
Phylogenetic relationships (UPGMA) of 22 species of Capsicum based on NADH dehydrogenase.

species like *C. eximium, C. coccineum, C. flexuosum,* and *C. hunzikerianum* had distinct branches and separated from other *Capsicum* taxa (**Figure 1**).

The phylogenetic study on the waxy gene [2] indicated that *C. chinense* and *C. frutescens* were grouped together. More so, a study on the trnC-rpoB intron, trnH-psbA intron, and waxy gene sequence data from seven *Capsicum spp*. also revealed that *C. chinense, C. annuum, and C. frutescens* grouped in the same cluster [2]. The above grouping supported our *Capsicum's* phylogenetic analysis based on NADH Dehydrogenase (**Figure 1**).

5. Genome organization

Genome sequence information of hot pepper revealed 37,989 scaffolds with an estimated size of 3.48 Gb [68]. The GC content was 35.03%, and there were 34,903 genes with an average exon and intron length of 286.5 pb and 541.6 bp, respectively [68]. These protein-coding genes of *Capsicum* were relatively the same as other Solanaceae species—tomato (34,771 genes) and potato (39,031 genes) [69–71].

The genetic maps of tomato and pepper are nearly comparable in length, with 1275 cm in tomato and 1246 cm in pepper [72]. However, it was determined that the average recombination rate/unit of physical distance in pepper and tomato is not the same [72]. This would happen if recombination were limited to homologous genes, as demonstrated in maize [73–83].

Regarding the number of homologous and segregating loci found by a probe, tomato, and pepper genomes differ, with pepper having a higher copy number [84]. The significant number of probes detecting multiple loci in pepper than in tomato might be related to the detection of more loci per probe in pepper, or it could be suggestive of a higher degree of interspecific polymorphism in *Capsicum* than in the interspecific cross used to generate the tomato map [84–86].

In plants, an increase in the amount of repetitive DNA has been identified as a cause of genome expansion [68, 87]. Retrotransposons distributed uniformly across gene-rich and gene-poor sections of the genome are likely to explain differences in nuclear DNA content among Solanaceae species [68, 81, 87]. Concerning transposable elements (TEs), *Capsicum's* TEs were predominantly composed of long terminal repeats (LTR). Specifically, most of the LTRs were Gypsy elements [68].

AFLP, SSR, RAPD, isozymes, inter-simple sequence repeat (ISSR), restrictions fragment length polymorphism (RFLP), gene-based makers, expressed sequence tag-simple sequence repeat (SCoT and EST SSR), and single nucleotide polymorphism (SNP) have all been used in the study and characterization of genetic diversity, phylogenetic relationships, genotypic variations, selection of parentals and progenies, cultivar identity, phenotypic characteristics, purity, population studies, and resistance to disease in *Capsicum* species [88, 89].

Chili peppers have been identified, and their germplasm diversity was evaluated using various molecular markers [90, 91]. Rodriguez discovered diagnostic RAPD (randomly amplified polymorphic DNA) producers for four domesticated species (including *C. chacoense*) but not for *C. frutescens* in a review [92]. Primarily, isozymes have been used to measure genetic diversity and define their genetic relationships within the genus [93].

Studies in the Solanaceae family linking the genetic maps of tomato and potato, respectively, sparked the discipline of comparative plant genomics in 1998 [94, 95]. The initial analysis discovered that the main difference between the tomato and potato genomes was paracentric inversions, and further investigations revealed that five inversions separated the two species [69, 86, 96]. It was shown that no map of *Capsicum* has yet been developed that accomplishes the aim of thoroughly defining and saturating the pepper chromosomes [72].

About 655 of the 1007 markers produced could be examined for divergence from single-locus Mendelian ratios [72]. Slightly more than half of the tested subgroup (337 = 50.7%) indicated variation from predicted ratios ($p = 0.01$), with p values as low as 2.69 × 10–25 in 81 of them (12.2%) [72].

The pepper tomato comparative map can be used in conjunction with the *Capsicum*, Lycopersicon, and Solanum phylogeny. It can also be used to identify conserved linkage blocks, reconstruct portions of the genome of these species' most recent ancestor, and, in some cases, determine lineage rearrangements that occurred [69, 71, 72, 97]. Because the number of ad hoc hypotheses utilizing either condition as the ancestral state is the same for pepper and tomato/potato, only two different arrangements can be provided [72]. It was noted that paracentric, as well as pericentric inversions and translocations, were the most common structural changes [72]. Interestingly, all tomato clones examined were hybridized to pepper DNA [72].

There were differences in tomato and pepper genomes regarding the amount of homologous and segregating loci, with pepper having a larger copy number [72]. The increased number of probes identifying multiple loci in pepper compared to tomato might be due to the detection of more loci per probe in pepper or indicative of a higher degree of interspecific polymorphism with *Capsicum* [72]. Regardless, the discrepancies in copy numbers across the specifics lacked the patterns consistent with systemic duplication [72].

Increases in the quantity of repetitive DNA have been identified as a source of genome extension in plants. A recent study in the Gramineae has revealed a pattern of retroelement increases between the genes of large-genome species compared to smaller-genome species analysis of repetitive DNA in the pepper genome indicated that 5% of the pepper genome was made up of elements with copy numbers >10,000, 26% with copy numbers >150, and 65% single-copy sequences [81, 85, 98–100]. The blocks of constitutive heterochromatin (7% of total karyotypic length) detected primarily at the telomers of *C. annum* cannot explain all of the additional DNA in pepper compared to tomato chromosomes [101]. As a result, differences in nuclear DNA content between tomato and pepper are likely to be explained by retrotransposons interspersed evenly across both gene-rich and gene-poor areas of the genome, as shown in the Gramineae [102].

The genetic maps of tomato and pepper are nearly comparable in length, with 1275 cm in tomato and 1246 cm in pepper [73–76]. Because of the comparable lengths of the genetic maps and the difference in DNA content, the recombination rate per unit of physical distance in pepper and tomato is not the same [73–76]. This would happen if recombination were limited to homologous genes, as predicted and demonstrated in maize [73–83].

A map of *Capsicum* has yet to be established that achieves the goal of properly identifying and saturating the pepper chromosomes [72]. Pepper has similar genetic content to tomato in the tomato-pepper investigation [72]. The fundamental difference between the tomato and potato genomes was identified as paracentric inversions, and subsequent research indicated that five inversions separated the two species [71, 72]. The tomato-pepper study also discovered that pepper and tomato had similar genomic content, as evidenced by the presence of pepper sequences that were complementary to all tomato cDNA tested [86]. However, the pepper genome had been significantly rearranged, with numerous pepper chromosomes that contain discrete tomato segments [72]. It was further discovered that the pepper genome had lost parts homologous to the tomato genome, but this did not change the fact that the homoeologous linkage blocks in the pepper genome had been considerably broken [72].

The number of homologous and segregating loci of tomato and pepper genomes differ, with pepper having a higher copy number [72]. The larger number of probes detecting multiple loci in pepper than in tomato might be related to the detection of more loci per probe in pepper, or it could be suggestive of a higher degree of interspecific polymorphism in *Capsicum* than in the interspecific cross used to generate the tomato map [72, 103]. The differences in copy quantity across the specificity, on the other hand, lack the patterns associated with systemic duplications [72].

In plants, an increase in the amount of repetitive DNA has been identified as a cause of genome expansion [72]. Retrotransposons are distributed uniformly across both gene-rich and gene-poor sections of the genome and may explain the differences in pepper and tomato nuclear DNA content [72]. Tomato and pepper genetic maps are approximately identical in length, with 1275 cm in tomato and 1246 cm in pepper [72]. The difference in the average recombination rate in pepper and tomato may be due to variable lengths of their genetic maps and differences in DNA content [72].

6. Genes of *Capsicum*

Annotated gene sequences are crucial for breeding and developing varieties for tolerance and resistance to biotic and abiotic stresses and enhancing Capsicum's agronomic and nutritional traits. NCBI record during the time of this writing showed a total of 1181 annotated Capsicum genes (duplicate copies are included). Specifically, *Capsicum chacoense* has *132 annotated genes, followed by Capsicum galapagoense* (132), *Capsicum eximium* (132), *Capsicum frutescens* (131), *Capsicum baccatum var. baccatum* (131), *Capsicum baccatum var. pendulum* (131), *Capsicum baccatum var. praetermissum* (131), *Capsicum pubescens* (131), and *Capsicum lycianthoides* (130).

Boswell [104] looked at the inheritance of 16 phenotypes in pepper and discovered seven gene symbols for purple foliage and stem color, blunt fruit apex, bulged fruit base, pendent fruit position, red mature fruit color, strong purple foliage and stem color, and non-clasping fruit calyx [104, 105]. Data was merged from six different maps from the US, Israel, and France to construct an integrated *Capsicum* genetic map with six distinct progenies and 2262 genetic markers spanning 1832 cm [105, 106].

The first *Capsicum* gene nomenclature and symbols were published in 1865 [105]. Lippert and colleagues increased the number of genes on the list to 75 [105]. Daskalov published a gene list in Bulgarian with around 90 genes [105]. Greenleaf developed a gene list for pepper breeding based on Lippert and others' gene lists and included several extra gene symbols for pepper breeders' use [105, 107, 108]. This gene list includes morphological features, physiological traits, sterility, resistance to diseases, nematodes, and herbicides among *Capsicum's* 292 known genes [105, 108]. The *Capsicum* and Eggplant Newsletter Editorial Board (CENL) proposed the criteria for *Capsicum* gene nomenclature in 1994 to help standardize and articulate the gene symbols [105]. A list of known genes using these rules, reallocating some gene symbols, and standardizing confusing symbols were compiled [105, 109]. Ninety-two genes have been added to Daskalov and Poulos' gene list [105, 109]. The suggested gene symbols complied with the *Capsicum* gene nomenclature guidelines for those features tested for inheritance according to CENL [105, 109]. An attempt was made to fix inaccuracies in earlier lists' gene symbols and descriptions [105].

Since Webber (1912) explored the inheritance of various phenotypes of *Capsicum* genes [105]. The inheritance of 16 features in pepper identifies seven gene symbols for seven distinct phenotypes [104]. Following Boswell's work, pepper inheritance research grew in popularity, with more essential features connected with the increased global significance of pepper production and more induced or spontaneous mutants being produced [104, 105].

More efforts to tag identical genes with molecular markers have been made since the late 1980s [110]. In addition, efforts were made to clone and describe the genes [110].

7. Applications

Medicinal herbs are essential natural medicines for various illnesses [111–114]. Higher plant natural products have the potential to provide a new reservoir of therapeutic medicines with unique modes of action [114–116]. Peppers were shown to have elevated amounts of vitamins C and E and provitamin A, carotenoids, and phenolic compounds, all of which contribute to the plant's overall antioxidant activity and bioactive qualities [117, 118]. Capsaicinoids (vanillylamine) coupled to a branched-chain fatty acid are the most common phenolic compounds discovered in pepper fruits [117–119]. Capsaicin and dihydrocapsaicin, for example, are responsible for 90% of pepper pungency [120]. The pungency of *Capsicum* varies depending on the species and cultivar. The concentrations of these chemicals can range from 0 mg/ 100 g in non-pungent cultivars to 664 mg/100 g in pungent cultivars [121]. *Capsicum spp*. also includes the capsinoids capsiate and dihydrocapsiate, two non-pungent analogs of capsaicin and dihydrocapsaicin, respectively [119]. Flavonol and flavone glycosides, as well as hydroxycinnamic acids, are other phenolic chemicals [119]. *Capsicum spp.* has many health advantages, and consumption of Capsicum is a part of a regular diet of diverse people of different ethnicities worldwide [119].

Due to bioactive chemicals, pepper (*C. annum* L.) has been reported to heal various degenerative human ailments [122, 123]. Bell peppers (sometimes called sweet peppers or peppers) are high in phenolic components such as quercetin, luteolin, and capsaicinoids [111, 124–126]. These phenolic chemicals protect against cancer, diabetes, oxidative stress, cardiovascular disease, and neurodegenerative illnesses, including Parkinson's and Alzheimer's [111, 123, 124].

Capsicum spp. are native to Mexico and Central America and have been utilized in traditional medicinal practices by Aztecs and Mayans from pre-Hispanic times [123]. The most notable is the indigenous medicine man Martin de la Cruz's Libellus de *Medicinabilus Indorum* Herbtis (Little Book of the Indians' Medicinal Herbs) [127]. *Capsicum spp*. were reported to have approximately 32 different health-related uses by the indigenous Mayan inhabitants of Mesoamerica at the turn of the twentieth century, including treatment for arthritis, rheumatism, stomach aches, skin rashes, and relief from dog and snake bites [128]. *On* the other hand, *Capsicum* fruits are not just used in Latin America; their medical benefits, as well as their use and production, have expanded worldwide [128]. As a result, *Capsicum* fruits are cited in the "Blue Beryll," a traditional Tibetan medical treatise, to improve the digestive warmth of the stomach and as a treatment for edema, hemorrhoids, parasitic protozoa, and leprosy [1, 128].

Furthermore, they are regarded as antispasmodic in Africa, disinfectants, antiirritants, and antitussive agents for the lungs [129]. *Capsicum* fruits are used topically for pain, neuropathy, cluster headaches, migraines, psoriasis, trigeminal neuralgia, and herpes zoster [129–131]. Dyspepsia, lack of appetite, flatulence, atherosclerosis, stroke, heart disease, and muscular tension have all been treated with it [132]. Today, a fifth of the world's population regularly consumes fresh or dried fruits as spices, food supplements, and additives [132].

Natural bioingredients are increasingly used in food for preservation, shelf life extension, and microbiological safety [111, 115]. Spices are used in a wide variety of meals due to the various phytochemicals they contain [111, 115]. Spices such as ginger, allspice, pepper, nutmeg, cloves, celery leaves, chives, and pepper are produced worldwide [111, 115]. Capsaicin, the main chemical ingredient in spicy peppers, has been shown to exhibit antibacterial action against Gram-negative and Gram-positive spoilage bacteria, as well as pathogenic bacteria [133].

Carotenoids are phytochemicals found in *Capsicum* that function as scavengers of singlet molecular oxygen, peroxyl radicals, and reactive nitrogen species (RNS), and they protect cells and tissues from reactive oxygen species (ROS) damage [134]. Despite this, *Capsicum spp.* has a large amount of total antioxidant activity. It is linked not only to its vitamin and carotenoid levels but also to its phenolic composition [117, 118]. *C. annum, C. frutescens,* and *C. chinense*, the antioxidant ingredients (carotenoids, flavonoids, phenolic acids, and ascorbic acid), rise in concentration along with the antioxidant activity measured *in vitro* as the fruit matures [135]. Furthermore, *Capsicum* has a more potent antioxidant activity than other veggies [135]. Capsaicin exhibits antioxidant properties similar to butylhydroxyanisole (BHA). It can protect human low-density lipoprotein (LDL) from oxidation, as well as block copper ion-induced lipid peroxidation, and reduce the development of thiobarbituric acid reactive substance (TBARS) [136, 137].

Capsaicin has received much attention because of its ability to cause apoptosis in various cancer cell lines, including pancreatic, colonic, prostatic, liver, esophageal, bladder, skin, leukemia, lung, and endothelium cells, while leaving normal cells unaffected [138]. However, because cancer prevention and promotion have been advocated, its role in carcinogenesis remains contentious [139]. The promotor effect appears to be linked to high consumption of capsaicin in the diet [139]. In this regard, a meta-analysis from 2014 recommended that capsaicin use should be modest [140].

In several investigations, *Capsicum* has various effects on glucose metabolism *in vitro* and *in vivo* [123, 140, 141]. Selected pungent and non-pungent *Capsicum* cultivars have demonstrated significant antioxidant activity as well as a great inhibitory profile on carbohydrate-degrading enzymes like—glycosidase, which is linked to glucose absorption [123, 140, 141].

Eating hot *Capsicum spp*. may improve postprandial glucose, insulin, and energy metabolism [85]. Other research has discovered that *C. chinense* (habanero) has a stronger anti-amylase activity than anti-glucosidase activity [85].

The digestive stimulatory activity of hot *Capsicum* is thought to be linked to the stimulation of saliva and bile production, as well as pancreatic and small intestine digestive enzyme activities [140–142]. It also boosts saliva production and salivary amylase activity, which aid in the digestion of starch and mucous membrane development in the mouth, throat, and gastrointestinal system [140, 141]. Hot *Capsicum* has been shown in animal experiments to improve fat digestion and absorption in high-fat-fed animals by stimulating the liver to release bile rich in bile acids [132].

Natural capsaicinoids from chili peppers have gotten much press as topical pain treatments [130, 131, 143, 144]. Capsaicin's unique ability has been used with lotions, ointments, and patches to treat a variety of pains, including neuropathic pain [130, 131, 145].

Chili peppers and their contents might be helpful and promising in preventing or treating insulin resistance, hypertension, dyslipidemia, and obesity [123, 140].

When tested in animal models, capsaicinoids and other bioactive chemicals from *Capsicum* appear to have additional health benefits [123, 140]. Supplementing with capsaicin may enhance physical activities such as grid, strength, and endurance performance by boosting live glycogen content [123, 140]. In animal trials, it can also reduce several exercise-induced tiredness indices [146, 147]. The use of spicy chilies in a regular diet has been linked to improved iron levels in the population [123, 140, 146]. In hamsters fed diets containing capsaicinoids, the capsaicinoids may lower total plasma cholesterol, inhibit the development of atherosclerotic plaque, and relax the aorta artery via increasing fecal excretion of acidic sterols [123, 140, 146]. It may also have a positive vascular function and modify plasma lipids [123, 140, 146]. Capsaicin, taken orally or topically, lowers rheumatoid arthritis pain, inflammatory heat, and unpleasant chemical hyperalgesia, according to research examining several therapies for knee osteoarthritis in elderly individuals [123, 130, 140, 146]. Capsaicin is claimed to alter 5-lipoxygenase, a major enzyme involved in the manufacture of the inflammatory mediators' leukotrienes, in human polymorphonuclear leukocyte cells [140, 141, 146, 148].

8. Conclusion

Capsicum is a well-diverse shrub commonly grown in different parts of the world. It is genetically diverse, with numerous cultivars and taxonomic derivatives. It has a widely described chromosome number of 2n = 24; however, some wild *Capsicum* species, such as *C. buforum, C. capylopodiume,* and *C. capylopodiume* have a ploidy level of 2n = 26. *C. frutescens* showed karyotypic formulae for 11M + 1SM + 1A and 11M + 1A. *Capsicum* has a relatively similar genomic content compared to tomato, as evidenced by the presence of Capsicum sequences complementary to many tomato cDNAs. To date, more than a thousand Capsicum genes have been annotated in the GenBank that will aid future research on improving the yield and biotic and abiotic traits of *Capsicum*. This comprehensive review is essential in understanding *Capsicum's* biology, genetics, and genomics toward improving its horticultural traits and nutritional and medicinal values.

Acknowledgements

The authors would like to thank the reviewers and colleagues who provided their valuable comments. Appreciation to the Department of Biology, College of Science and Technology, Adventist University of the Philippines, for the support.

Conflict of interest

The authors declare no conflict of interest.

Author details

Nkwiza M. Nankolongo[1], Orlex Baylen Yllano[1,5*], Leilani D. Arce[2], Neil John V. Vegafria[2], Ephraim A. Evangelista[3], Ferdinand A. Esplana[3], Lester Harris R. Catolico[4], Merbeth Christine L. Pedro[4] and Edgar E. Tubilag[4]

1 Department of Biology, College of Science and Technology, Philippines

2 Botany and Systematics Laboratory, Department of Biology, CST, Adventist University of the Philippines, Philippines

3 Microbiology Laboratory, Department of Biology, CST, Adventist University of the Philippines, Philippines

4 Anatomy and Physiology Laboratory, Department of Biology, CST, Adventist University of the Philippines, Philippines

5 Cell and Molecular Biology Laboratory, Department of Biology, CST, Adventist University of the Philippines, Silang Cavite, Philippines

*Address all correspondence to: obyllano@aup.edu.ph

References

[1] Basu SK, De AK. Capsicum: Historical and botanical perspectives. 2003. Available from: https://www.semanticscholar.org/paper/Capsicum%3A-historical-and-botanical-perspectives-Basu-De/b1d8db5673dca8e98bed971f4a8e7ec0254fea9f

[2] Jarret RL. DNA barcoding in a crop genebank: The Capsicum annuum species complex. The Open Biology Journal. 2008;**1**(1):35-42

[3] Aguilar-Melendez A, Morrell PL, Roose ML, Kim S-C. Genetic diversity and structure in semiwild and domesticated chiles (*Capsicum annuum*; Solanaceae) from Mexico. American Journal of Botany. 2009;**96**(6):1190-1202. DOI: 10.3732/ajb.0800155

[4] Hunziker AT, Hawkes JG, Lester RN, Skelding AD. The biology and taxonomy of the Solanaceae. In: Linnean Society Symposium Series, London; 1979;**7**:49-85

[5] Invasive Species Compendium, *Capsicum annum*. 2022. Available from: https://www.cabi.org/isc/datasheet/15784#tosummaryOfInvasiveness

[6] Luna-Ruiz J d J, Nabhan GP, Aguilar-Meléndez A. Shifts in plant chemical defenses of Chile pepper (*Capsicum annuum* L.) due to domestication in Mesoamerica. Frontiers in Ecology and Evolution. 2018;**6**:48. DOI: 10.3389/fevo.2018.00048

[7] FAO EcoCrop. *Capsicum annuum* L, eco-crop online database. In: Land and Water Development Division. Food and Agricultural Organization of the UN (FAO); 2014. Available from: http://ecocrop.fao.org/ecocrop/srv/en/dataSheet?id=618

[8] Liogier HA, Martorell LF. Flora of Puerto Rico and Adjacent Islands: A Systematic Synopsis, 2nd Edition Revised. San Juan, Puerto Rico: La Editorial, University of Puerto Rico; 2000. p. 382

[9] Randall RP. A global compendium of weeds. 3rd Ed. Perth. 2017:3653

[10] Vascular Plants of Ecuador. Catalogue of the Vascular Plants of Ecuador, Tropicos Website. St. Louis, Missouri and Cambridge, Massachusetts, USA: Missouri Botanical Garden and Harvard University Herbaria; 2014. Available from: http://tropicos.org/Project/CE

[11] Pickersgill B. Cytogenetics and evolution of Capsicum L. Chromosome engineering in plants: Genetics, breeding, evolution, part B. Amsterdam: Elsevier; 1991:139-160

[12] Tong N, Bosland PW. No title found. Genetic Resources and Crop Evolution. 2003;**50**(2):193-199. DOI: 10.1023/A:1022986615694

[13] Lanteri S, Pickersgill B. Chromosomal structural changes in *Capsicum annuum* L. and C. chinense Jacq. Euphytica. 1993;**67**(1-2):155-160. DOI: 10.1007/BF00022739

[14] Moscone EA. Analysis of nuclear DNA content in Capsicum (Solanaceae) by flow cytometry and Feulgen densitometry. Annals of Botany. 2003;**92**(1):21-29. DOI: 10.1093/aob/mcg105

[15] Kang BC, Nahm SH, Huh JH, Yoo HS, Yu JW, Lee MH, et al. An interspecific (*Capsicum annuum* × *C. chinese*) F2 linkage map in pepper using RFLP and AFLP markers. Theoretical and Applied Genetics. 2001;**102**(4):531-539. DOI: 10.1007/s001220051678

[16] Schreiber ME. Cloning and characterization of repetitive DNA in Pepper (Capsicum annuum) [M.S. thesis]. Fresno: California State University; 2004. p. 49

[17] Almeida BM de, Martins L do V, Lopes ÂC de A, Gomes RLF, Valente SE dos S, Peron AP, Silva VB da, Feitoza L de L. Karyotype polymorphism of GC-rich constitutive heterochromatin in Capsicum L. pepper accessions Crop Breeding and Applied Biotechnology 2022;**22**(1):e38642113. DOI: 10.1590/1984-70332022v22n1a03

[18] Reed DH, Frankham R. Correlation between fitness and genetic diversity. Conservation Biology. 2003;**17**(1):230-237. DOI: 10.1046/j.1523-1739.2003.01236.x

[19] Souza WR do N, Almeida AC de, Carvalho RD, Ferreira RL, Peron AP. Karyotypic characterization of Capsicum sp. Accessions. Acta Scientiarum. Agronomy. 2015;**37**(2):147. DOI: 10.4025/actasciagron.v37i2.19485

[20] Souza SAM, Martins KC, Pereira TNS. Polimorfismo cromossômico em Capsicum chinense Jacq. Ciência Rural. 2011;**41**(10):1777-1783. DOI: 10.1590/S0103-84782011001000017

[21] Carvalho SI de, Bianchetti L de B, Reifschneider FJ. Registro e proteção de cultivares pelo setor público: A experiência do programa de melhoramento de Capsicum da Embrapa Hortaliças. Horticultura Brasileira, 2009;**27**(2):135-138. DOI: 10.1590/S0102-05362009000200002

[22] Guerra N. Estudios cromossómicos de cuatro seleciones de Capsicum chinense Jacq. Revista Científica UDO Agrícola. 2001;**1**(1):34-41. Available from: https://dialnet.unirioja.es/servlet/articulo?codigo=2221473

[23] Pickersgill B. No title found. Euphytica. 1997;**96**(1):129-133. DOI: 10.1023/A:1002913228101

[24] Pozzobon MT, Schifino-Wittmann MT. A meiotic study of the wild and semi-domesticated Brazilian species of genus Capsicum L. (Solanaceae). Cytologia. 2006;**71**(3):275-287. DOI: 10.1508/cytologia.71.275

[25] Teodoro-Pardo CV, García-Velázquez A, Corona-Torres T. Polimorfismo cromossômico en *Capsicum annuum* L.(Solanaceae) em recolectas de Puebla, Morelos y Quartenário, México. Agrociencia. 2007;**41**(10):873-881. Available from: https://agrociencia-colpos.mx/index.php/agrociencia/article/view/594/594

[26] Pickersgill B. Domestication of plants in the Americas: Insights from Mendelian and molecular genetics. Annals of Botany. 2007;**100**(5):925-940. DOI: 10.1093/aob/mcm193

[27] Moscone EA, Lambrou M, Ehrendorfer F. Fluorescent chromosome banding in the cultivated species of Capsicum (Solanaceae). Plant Systematics and Evolution. 1996;**202**(1-2):37-63. DOI: 10.1007/BF00985817

[28] Wadt LH d O, Ehringhaus C, Kageyama PY. Genetic diversity of "Pimenta Longa" genotypes (Piper spp., Piperaceae) of the Embrapa acre germplasm collection. Genetics and Molecular Biology. 2004;**27**(1):74-82. DOI: 10.1590/S1415-47572004000100013

[29] Zarco CR. A new method for estimating karyotype asymmetry. Taxon. 1986;**35**(3):526-530. DOI: 10.2307/1221906

[30] Day PR. Managing Global Genetic Resources: Agricultural Crop Issues and Policies. Washington: National Academy Press; 1993:449

[31] van Zonneveld M, Ramirez M, Williams DE, Petz M, Meckelmann S, Avila T, et al. Screening genetic resources of Capsicum peppers in their primary Center of Diversity in Bolivia and Peru. PLoS One. 2015;**10**(9):e0134663. DOI: 10.1371/journal.pone.0134663

[32] Leimu R, Mutikainen PIA, Koricheva J, Fischer M. How general are positive relationships between plant population size, fitness and genetic variation? Journal of Ecology. 2006;**94**(5):942-952

[33] Rodríguez GR, Moyseenko JB, Robbins MD, Huarachi Morejón N, Francis DM, van der Knaap E. Tomato analyzer: A useful software application to collect accurate and detailed morphological and colorimetric data from two-dimensional objects. Journal of Visualized Experiments. 2010;**37**:1856. DOI: 10.3791/1856

[34] Lozada DN, Bhatta M, Coon D, Bosland PW. Single nucleotide polymorphisms reveal genetic diversity in new Mexican Chile peppers (Capsicum spp.). BMC Genomics. 2021;**22**(1):356. DOI: 10.1186/s12864-021-07662-7

[35] Espichán F, Rojas R, Quispe F, Cabanac G, Marti G. Metabolomic characterization of 5 native Peruvian chili peppers (Capsicum spp.) as a tool for species discrimination. Food Chemistry. 2022;**386**:132704. DOI: 10.1016/j.foodchem.2022.132704

[36] Taher D, Solberg SØ, Prohens J, Chou Y, Rakha M, Wu T. World vegetable Center eggplant collection: Origin, composition, seed dissemination and utilization in breeding. Frontiers in Plant Science. 2017;8:1484. DOI: 10.3389/fpls.2017.01484

[37] Passam H, Karapanos I. Eggplants, peppers and tomatoes: Factors affecting the quality and storage life of fresh and fresh-cut (minimally processed) produce. The European Journal of Plant Science and Biotechnology. 2016;**2**:156-170

[38] Doganlar S, Frary A, Daunay M-C, Lester RN, Tanksley SD. Conservation of gene function in the Solanaceae as revealed by comparative mapping of domestication traits in eggplant. Genetics. 2002;**161**(4):1713-1726. DOI: 10.1093/genetics/161.4.1713

[39] Zygier S, Chaim AB, Efrati A, Kaluzky G, Borovsky Y, Paran I. QTLs mapping for fruit size and shape in chromosomes 2 and 4 in pepper and a comparison of the pepper QTL map with that of tomato. Theoretical and Applied Genetics. 2005;**111**(3):437-445. DOI: 10.1007/s00122-005-2015-7

[40] Hurtado M, Vilanova S, Plazas M, Gramazio P, Herraiz FJ, Andújar I, et al. Phenomics of fruit shape in eggplant (Solanum melongena L.) using tomato Analyzer software. Scientia Horticulturae. 2013;**164**:625-632. DOI: 10.1016/j.scienta.2013.10.028

[41] Brewer MT, Lang L, Fujimura K, Dujmovic N, Gray S, van der Knaap E. Development of a controlled vocabulary and software application to analyze fruit shape variation in tomatoes and other plant species. Plant Physiology. 2006;**141**(1):15-25. DOI: 10.1104/pp.106.077867

[42] Barchi L, Rabanus-Wallace MT, Prohens J, Toppino L, Padmarasu S, Portis E, et al. Improved genome assembly and pan-genome provide key insights into eggplant domestication and breeding. The Plant Journal. 2021;**107**(2):579-596. DOI: 10.1111/tpj.15313

[43] Nunome T, Ishiguro K, Yoshida T, Hirai M. Mapping of fruit shape and color

development traits in eggplant (*Solanum melongena* L.) based on RAPD and AFLP markers. Breeding Science. 2001;**51**(1):19-26. DOI: 10.1270/jsbbs.51.19

[44] Knapp S. Tobacco to tomatoes: A phylogenetic perspective on fruit diversity in the Solanaceae. Journal of Experimental Botany. 2002;**53**(377):2001-2022. DOI: 10.1093/jxb/erf068

[45] García-González CA, Silvar C. Phytochemical assessment of native Ecuadorian peppers (Capsicum spp.) and correlation analysis to fruit Phenomics. Plants. 2020;**9**(8):986. DOI: 10.3390/plants9080986

[46] Rodríguez GR, Muños S, Anderson C, Sim S-C, Michel A, Causse M, et al. Distribution of *SUN, OVATE, LC*, and *FAS* in the tomato germplasm and the relationship to fruit shape diversity. Plant Physiology. 2011;**156**(1):275-285. DOI: 10.1104/pp.110.167577

[47] Paran I, van der Knaap E. Genetic and molecular regulation of fruit and plant domestication traits in tomato and pepper. Journal of Experimental Botany. 2007;**58**(14):3841-3852. DOI: 10.1093/jxb/erm257

[48] Naegele RP, Mitchell J, Hausbeck MK. Genetic diversity, population structure, and heritability of fruit traits in Capsicum annuum. PLoS One. 2016;**11**(7):e0156969. DOI: 10.1371/journal.pone.0156969

[49] Rehman F, Gong H, Li Z, Zeng S, Yang T, Ai P, et al. Identification of fruit size associated quantitative trait loci featuring SLAF-based high-density linkage map of goji berry (Lycium spp.). BMC Plant Biology. 2020;**20**(1):474. DOI: 10.1186/s12870-020-02567-1

[50] Chaim A, Borovsky Y, Rao G, Gur A, Zamir D, Paran I. Comparative QTL mapping of fruit size and shape in tomato and pepper. Israel Journal of Plant Sciences. 2006;**54**(3):191-203. DOI: 10.1560/IJPS_54_3_191

[51] Adhikari P, McNellie J, Panthee DR. Detection of quantitative trait loci (QTL) associated with the fruit morphology of tomato. Genes. 2020;*11*(10):1117. DOI: 10.3390/genes11101117

[52] Vilarinho LBO, Henriques da Silva DJ, Greene A, Salazar KD, Alves C, Eveleth M, et al. Inheritance of fruit traits in *Capsicum annuum*: Heirloom cultivars as sources of quality parameters relating to pericarp shape, color, thickness, and Total soluble solids. Journal of the American Society for Horticultural Science. 2015;**140**(6):597-604. DOI: 10.21273/JASHS.140.6.597

[53] MoulinMM, RodriguesR, RamosHCC, Bento CS, Sudré CP, Gonçalves LSA, et al. Construction of an integrated genetic map for *Capsicum baccatum* L. Genetics and Molecular Research. 2015;**14**(2):6683-6694. DOI: 10.4238/2015.June.18.12

[54] Bernacchi D, Tanksley SD. An interspecific backcross of *Lycopersicon esculentum* × *L. hirsutum*: Linkage analysis and a QTL study of sexual compatibility factors and floral traits. Genetics. 1997;**147**(2):861-877. DOI: 10.1093/genetics/147.2.861

[55] Sage-Palloix AM, Jourdan F, Phaly T, Nemouchi G, Lefebvre V, Palloix A. Analysis of diversity in pepper genetic resources: Distribution of horticultural and resistance traits in the INRA pepper germplasm. In: Progress in research on capsicum and eggplant. Proceedings of the XIIIth EUCARPIA Meeting, Warsaw. 2007:33-42

[56] Ashrafi H, Kinkade M, Foolad MR. A new genetic linkage map of tomato

based on a *Solanum lycopersicum* × *S. pimpinellifolium* RIL population displaying locations of candidate pathogen response genes. Genome. 2009;**52**(11):935-956. DOI: 10.1139/G09-065

[57] Stall RE, Jones JB, Minsavage GV. Durability of resistance in tomato and pepper to Xanthomonads causing bacterial spot. Annual Review of Phytopathology. 2009;**47**(1):265-284. DOI: 10.1146/annurev-phyto-080508-081752

[58] Haggard JE, Johnson EB, St. Clair, D. A. Multiple QTL for horticultural traits and quantitative resistance to *Phytophthora infestans* linked on *Solanum habrochaites* chromosome 11. G3 Genes|Genomes|Genetics. 2015;**5**(2):219-233. DOI: 10.1534/g3.114.014654

[59] Nowicki M, Foolad MR, Nowakowska M, Kozik EU. Potato and tomato late blight caused by Phytophthora infestans: An overview of pathology and resistance breeding. Plant Disease. 2012;**96**(1):4-17

[60] Saei A, Hoeata K, Krebs A, Sutton P, Herrick J, Wood M, et al. The status of *Pseudomonas syringae* pv. *Actinidiae* (Psa) in the New Zealand kiwifruit breeding programme in relation to ploidy level. In IX International Symposium on Kiwifruit. 2017;**1218**:293-298

[61] Ifuku K, Endo T, Shikanai T, Aro E-M. Structure of the chloroplast NADH dehydrogenase-like complex: Nomenclature for nuclear-encoded subunits. Plant and Cell Physiology. 2011;**52**(9):1560-1568. DOI: 10.1093/pcp/pcr098

[62] Kraft KH, Brown CH, Nabhan GP, Luedeling E, Luna Ruiz J d J, Coppens d'Eeckenbrugge G, et al. Multiple lines of evidence for the origin of domesticated chili pepper, Capsicum annuum, in Mexico. Proceedings of the National Academy of Sciences. 2014;**111**(17):6165-6170. DOI: 10.1073/pnas.1308933111

[63] Bharath SM, Cilas C, Umaharan P. Fruit trait variation in a Caribbean germplasm collection of aromatic hot peppers (*Capsicum chinense* Jacq.). HortScience. 2013;**48**(5):531-538. DOI: 10.21273/HORTSCI.48.5.531

[64] Nicolaï M, Cantet M, Lefebvre V, Sage-Palloix A-M, Palloix A. Genotyping a large collection of pepper (Capsicum spp.) with SSR loci brings new evidence for the wild origin of cultivated *C. annuum* and the structuring of genetic diversity by human selection of cultivar types. Genetic Resources and Crop Evolution. 2013;**60**(8):2375-2390. DOI: 10.1007/s10722-013-0006-0

[65] Barboza GE, Carrizo García C, Leiva González S, Scaldaferro M, Reyes X. Four new species of Capsicum (Solanaceae) from the tropical Andes and an update on the phylogeny of the genus. PLoS One. 2019;**14**(1):e0209792. DOI: 10.1371/journal.pone.0209792

[66] Bento CDS, Sudré CP, Rodrigues R, Riva EM, Pereira MG. Descritores qualitativos e Multicategóricos na estimativa da variabilidade fenotípica entre acessos de pimentas. Scientia Agrária. 2007;**8**(2):149. DOI: 10.5380/rsa.v8i2.8379

[67] Carrizo García C, Barfuss MHJ, Sehr EM, Barboza GE, Samuel R, Moscone EA, et al. Phylogenetic relationships, diversification, and expansion of chili peppers (Capsicum, Solanaceae). Annals of Botany. 2016;**118**(1):35-51. DOI: 10.1093/aob/mcw079

[68] Dooner HK. Genetic fine structure of the bronze locus in maize. Genetics. 1986;**113**(4):1021-1036. DOI: 10.1093/genetics/113.4.1021

[69] Tomato Genome Consortium. The tomato genome sequence provides insights into fleshy fruit evolution. Nature. 2012;**485**(7400):635

[70] Weeden NF, Muehlbauer FJ, Ladizinsky G. Extensive conservation of linkage relationships between pea and lentil genetic maps. Journal of Heredity. 1992;**83**(2):123-129. DOI: 10.1093/oxfordjournals.jhered.a111171

[71] Spooner DM, Anderson GJ, Jansen RK. Chloroplast DNA evidence for the interrelationships of tomatoes, potatoes, and PEPINOS (SOLANACEAE). American Journal of Botany. 1993;**80**(6):676-688. DOI: 10.1002/j.1537-2197.1993.tb15238.x

[72] Livingstone KD, Lackney VK, Blauth JR, van Wijk R, Jahn MK. Genome mapping in Capsicum and the evolution of genome structure in the Solanaceae. Genetics. 1999;**152**(3):1183-1202. DOI: 10.1093/genetics/152.3.1183

[73] Thuriaux P. Is recombination confined to structural genes on the eukaryotic genome? Nature. 1977;**268**(5619):460-462. DOI: 10.1038/268460a0

[74] Dooner HK, Weck E, Adams S, Ralston E, Favreau M, English J. A molecular genetic analysis of insertions in the bronze locus in maize. Molecular and General Genetics MGG. 1985;**200**(2):240-246. DOI: 10.1007/BF00425430

[75] Brown J, Sundaresan V. A recombination hotspot in the maize A1 intragenic region. Theoretical and Applied Genetics. 1991;**81**(2):185-188. DOI: 10.1007/BF00215721

[76] Civardi L, Xia Y, Edwards KJ, Schnable PS, Nikolau BJ. The relationship between genetic and physical distances in the cloned a1-sh2 interval of the *Zea mays* L. genome. Proceedings of the National Academy of Sciences. 1994;**91**(17):8268-8272. DOI: 10.1073/pnas.91.17.8268

[77] Eggleston WB, Alleman M, Kermicle JL. Molecular organization and germinal instability of R-stippled maize. Genetics. 1995;**141**(1):347-360. DOI: 10.1093/genetics/141.1.347

[78] Patterson GI, Kubo KM, Shroyer T, Chandler VL. Sequences required for paramutation of the maize b gene map to a region containing the promoter and upstream sequences. Genetics. 1995;**140**(4):1389-1406. DOI: 10.1093/genetics/140.4.1389

[79] Dooner HK, Martínez-Férez IM. Recombination occurs uniformly within the bronze gene, a meiotic recombination hotspot in the maize genome. The Plant Cell. 1997;**9**(9):1633-1646. DOI: 10.1105/tpc.9.9.1633

[80] Park M, Jo S, Kwon J-K, Park J, Ahn JH, Kim S, et al. Comparative analysis of pepper and tomato reveals euchromatin expansion of pepper genome caused by differential accumulation of Ty3/Gypsy-like elements. BMC Genomics. 2011;**12**(1):85. DOI: 10.1186/1471-2164-12-85

[81] SanMiguel P, Tikhonov A, Jin Y-K, Motchoulskaia N, Zakharov D, Melake-Berhan A, et al. Nested retrotransposons in the intergenic regions of the maize genome. Science. 1996;**274**(5288):765-768. DOI: 10.1126/science.274.5288.765

[82] Bennetzen JL, SanMiguel P, Chen M, Tikhonov A, Francki M, Avramova Z. Grass genomes. Proceedings of the National Academy of Sciences. 1998;**95**(5):1975-1978. DOI: 10.1073/pnas.95.5.1975

[83] Gale MD, Devos KM. Comparative genetics in the grasses. Proceedings of the National Academy of Sciences. 1998;**95**(5):1971-1974. DOI: 10.1073/pnas.95.5.1971

[84] Igwe DO, Afiukwa CA, Acquaah G, Ude GN. Genetic diversity and structure of *Capsicum annuum* as revealed by start codon targeted and directed amplified minisatellite DNA markers. Hereditas. 2019;**156**(1):32. DOI: 10.1186/s41065-019-0108-6

[85] Flavell RB, Bennett MD, Smith JB, Smith DB. Genome size and the proportion of repeated nucleotide sequence DNA in plants. Biochemical Genetics. 1974;**12**(4):257-269. DOI: 10.1007/BF00485947

[86] Prince JP, Pochard E, Tanksley SD. Construction of a molecular linkage map of pepper and a comparison of synteny with tomato. Genome. 1993;**36**(3):404-417. DOI: 10.1139/g93-056

[87] Brown A, Yllano OB, Arce LD, Evangelista EA, Esplana FA, Catolico LHR, et al. Characterization, comparative, and phylogenetic analyses of retrotransposons in diverse plant genomes. In: Genetic Polymorphisms-New Insights. London, UK, London, UK: IntechOpen; 2021

[88] Rodriguez JM, Berke T, Engle L, Nienhuis J. Variation among and within Capsicum species revealed by RAPD markers. Theoretical and Applied Genetics. 1999;**99**(1-2):147-156. DOI: 10.1007/s001220051219

[89] Ganal MW, Lapitan NLV, Tanksley SD. A molecular and cytogenetic survey of major repeated DNA sequences in tomato (*Lycopersicon esculentum*). Molecular and General Genetics MGG. 1988;**213**(2-3):262-268. DOI: 10.1007/BF00339590

[90] Lefebvre V. Molecular markers for genetics and breeding: Development and use in pepper (Capsicum spp.). In: Lörz H, Wenzel G, editors. Molecular Marker Systems in Plant Breeding and Crop Improvement Biotechnology in Agriculture and Forestry. Berlin: Springer; 2005. pp. 189-214

[91] Bennetzen JL, Freeling M. The unified grass genome: Synergy in Synteny. Genome Research. 1997;**7**(4):301-306. DOI: 10.1101/gr.7.4.301

[92] Eshbaugh WH. Peppers: History and exploitation of a serendipitous new crop discovery. In: Janick J, Simon JE, editors. New Crops. New York: John Wiley & Sons; 1993. pp. 132-139. Available from: https://hort.purdue.edu/newcrop/proceedings1993/v2-132.html

[93] Bonierbale MW, Plaisted RL, Tanksley SD. RFLP maps based on a common set of clones reveal modes of chromosomal evolution in potato and tomato. Genetics. 1988;**120**(4):1095-1103. DOI: 10.1093/genetics/120.4.1095

[94] Wu F, Tanksley SD. Chromosomal evolution in the plant family Solanaceae. BMC Genomics. 2010;**11**(1):182. DOI: 10.1186/1471-2164-11-182

[95] Gebhardt C, Ritter E, Barone A, Debener T, Walkemeier B, Schachtschabel U, et al. RFLP maps of potato and their alignment with the homoeologous tomato genome. Theoretical and Applied Genetics. 1991;**83**(1):49-57. DOI: 10.1007/BF00229225

[96] Diambra LA. Genome sequence and analysis of the tuber crop potato. Nature. 2011;**475**

[97] Gale MD, Devos KM. Plant comparative genetics after 10 years. Science. 1998b;**282**(5389):656-659. DOI: 10.1126/science.282.5389.656

[98] Chen M, SanMiguel P, de Oliveira AC, Woo S-S, Zhang H, Wing RA, et al. Microcolinearity in sh2 -homologous regions of the maize, rice, and sorghum genomes. Proceedings of the National Academy of Sciences. 1997;**94**(7):3431-3435. DOI: 10.1073/pnas.94.7.3431

[99] Panstruga R. A contiguous 60 kb genomic stretch from barley reveals molecular evidence for gene islands in a monocot genome. Nucleic Acids Research. 1998;**26**(4):1056-1062. DOI: 10.1093/nar/26.4.1056

[100] An CS, Kim SC, Go SL. Analysis of red pepper (*Capsicum annuum*) genome. Journal of Plant Biology. 1996;**39**(1):57-61. Available from: https://agris.fao.org/agris-search/search.do?recordID=KR19970002381

[101] Moscone EA, Lambrou M, Hunziker AT, Ehrendorfer F. Giemsa C-banded karyotypes in Capsicum (Solanaceae). Plant Systematics and Evolution. 1993;**186**(3-4):213-229. DOI: 10.1007/BF00940799

[102] Barakat A, Carels N, Bernardi G. The distribution of genes in the genomes of Gramineae. Proceedings of the National Academy of Sciences. 1997;**94**(13):6857-6861. DOI: 10.1073/pnas.94.13.6857

[103] Menancio-Hautea D, Fatokun CA, Kumar L, Danesh D, Young ND. Comparative genome analysis of mungbean (*Vigna radiata* L. Wilczek) and cowpea (*V. unguiculata* L. Walpers) using RFLP mapping data. Theoretical and Applied Genetics. 1993;**86**(7):797-810. DOI: 10.1007/BF00212605

[104] Boswell VR. Improvement and genetics of tomatoes, peppers and eggplant. Improvement and genetics of tomatoes, peppers and eggplant. USA: United States Department of Agriculture; 1937:177-206

[105] Deyan W, Paul BW. The genes of Capsicum. HortScience. 2006;**41**(5):1169-1187. Available from: https://journals.ashs.org/hortsci/downloadpdf/journals/hortsci/41/5/article-p1169.xml

[106] Paran I, van der Voort JR, Lefebvre V, Jahn M, Landry L, van Schriek M, et al. An integrated genetic linkage map of pepper (Capsicum spp.). Molecular Breeding. 2004;**13**(3):251-261. DOI: 10.1023/B:MOLB.0000022526.30914.31

[107] Bassett MJ. Breeding vegetable crops. AVI Pub. Co. Westport; 1986

[108] Daskalov S. Gene list for the pepper. Genetics and Plant Breeding. 1973a;**6**:401-408

[109] Daskalov S, Poulos JM. Updated Capsicum gene list. Capsicum Eggplant NSLW. 1994;**13**:16-26

[110] Lefebvre V, Palloix A, Caranta C, Pochard E. Construction of an intraspecific integrated linkage map of pepper using molecular markers and doubled-haploid progenies. Genome. 1995;**38**(1):112-121. DOI: 10.1139/g95-014

[111] Sharifi-Rad M, Mnayer D, Morais-Braga MFB, Carneiro JNP, Bezerra CF, Coutinho HDM, et al. *Echinacea* plants as antioxidant and antibacterial agents: From traditional medicine to biotechnological applications: Echinacea plants as antioxidant and antibacterial agents. Phytotherapy Research. 2018;**32**(9):1653-1663. DOI: 10.1002/ptr.6101

[112] Bagheri G, Mirzaei M, Mehrabi R, Sharifi-Rad J. Cytotoxic and antioxidant activities of *Alstonia scholaris*, *Alstonia venenata*, and *Moringa oleifera* plants from India. Jundishapur Journal of Natural Pharmaceutical Products. 2016;**11**(3):1-11. DOI: 10.17795/jjnpp-31129

[113] Stojanović-Radić Z, Pejcić M, Stojanović N, Sharifi-Rad J, Stanković N. Potential of *Ocimum basilicum* L. and *Salvia officinalis* L. essential oils against biofilms of *P. aeruginosa* clinical isolates. Cellular and Molecular Biology (Noisy-le-Grand, France). 2016;**62**(9):27-33

[114] Salehi B, Hernández-Álvarez AJ, del Mar Contreras M, et al. Potential Phytopharmacy and Food Applications of Capsicum spp.: A Comprehensive Review. Natural Product Communications. 2018;**13**(11):1543-1556. DOI: 10.1177/1934578X1801301133

[115] Sharifi-Rad J, Salehi B, Schnitzler P, Ayatollahi SA, Kobarfard F, Fathi M, et al. Susceptibility of herpes simplex virus type 1 to monoterpenes thymol, carvacrol, p-cymene and essential oils of *Sinapis arvensis* L., Lallemantia royleana Benth. And Pulicaria vulgaris Gaertn. Cellular and Molecular Biology. 2017;**63**(8):42-47. DOI: 10.14715/cmb/2017.63.8.10

[116] Sharifi-Rad J, Soufi L, Ayatollahi SAM, Iriti M, Sharifi-Rad M, Varoni EM, et al. Antibacterial effect of essential oil from Xanthium strumarium against Shiga toxin-producing Escherichia coli. Cellular and Molecular Biology (Noisy-le-Grand, France). 2016;**62**(9):69-74

[117] Parry C, Wang Y-W, Lin S, Barchenger DW. Reproductive compatibility in Capsicum is not necessarily reflected in genetic or phenotypic similarity between species complexes. PLoS One. 2021;**16**(3):e0243689. DOI: 10.1371/journal.pone.0243689

[118] Morales-Soto A, Gómez-Caravaca AM, García-Salas P, Segura-Carretero A, Fernández-Gutiérrez A. High-performance liquid chromatography coupled to diode array and electrospray time-of-flight mass spectrometry detectors for a comprehensive characterization of phenolic and other polar compounds in three pepper (*Capsicum annuum* L.) samples. Food Research International. 2013;**51**(2):977-984. DOI: 10.1016/j.foodres.2013.02.022

[119] Howard LR, Wildman RE. Antioxidant vitamin and phytochemical content of fresh and processed pepper fruit (*Capsicum annuum*). In: Wildman REC, editor. Nutraceuticals and Functional Foods. 2nd ed. Boca Raton, Florida: CRC Press; 2007. pp. 165-191

[120] Hamed M, Kalita D, Bartolo ME, Jayanty SS. Capsaicinoids, polyphenols and antioxidant activities of *Capsicum annuum*: Comparative study of the effect of ripening stage and cooking methods. Antioxidants. 2019;**8**(9):364. DOI: 10.3390/antiox8090364

[121] Ammar S, del Mar Contreras M, Belguith-Hadrich O, Segura-Carretero A, Bouaziz M. Assessment of the distribution of phenolic compounds and contribution to the antioxidant activity in Tunisian fig leaves, fruits, skins, and pulps using mass spectrometry-based analysis. Food & Function. 2015;**6**(12):3663-3677. DOI: 10.1039/C5FO00837A

[122] Chen C-Y, Yeh Y-T, Yang W-L. Amides from the stem of *Capsicum annuum*. Natural Product Communications. 2011;**6**(2):1934578X1100600. DOI: 10.1177/1934578X1100600217

[123] Salehi B, Hernández-Álvarez AJ, del Mar Contreras M, Martorell M, Ramírez-Alarcón K, Melgar-Lalanne G, et al. Potential phytopharmacy and food applications of *Capsicum* spp.: A comprehensive review. Natural Product Communications. 2018;*13*(11):1934578X1801301. DOI: 10.1177/1934578X1801301133

[124] Oboh G, Rocha JBT. Distribution and antioxidant activity of polyphenols

in ripe and unripe tree pepper (*Capsicum pubescens*). Journal of Food Biochemistry. 2007;**31**(4):456-473. DOI: 10.1111/j.1745-4514.2007.00123.x

[125] Liu Y, Nair MG. Capsaicinoids in the hottest pepper Bhut Jolokia and its antioxidant and anti-inflammatory activities. Natural Product Communications. 2010;**5**(1):91-94

[126] Si W, Man SW, Chenb Z-Y, Chung HY. Stability of capsaicinoid content at raised temperatures. Natural Product Communications. 2014;**9**(7):985-988

[127] Domínguez F, Alonso-Castro AJ, Anaya M, González-Trujano ME, Salgado-Ceballos H, Orozco-Suárez S. Mexican traditional medicine: Traditions of yesterday and phytomedicines for tomorrow. In: Duarte MCT, Rai M, editors. Therapeutic Medicinal Plants: From Lab to the Market. Boca Raton, Florida: CRC Press; 2015. pp. 10-46

[128] Cichewicz RH, Thorpe PA. The antimicrobial properties of Chile peppers (Capsicum species) and their uses in Mayan medicine. Journal of Ethnopharmacology. 1996;**52**(2):61-70. DOI: 10.1016/0378-8741(96)01384-0

[129] Elujoba A, Odeleye O, Ogunyemi C. Traditional medicine development for medical and dental primary health care delivery system in Africa. African Journal of Traditional, Complementary and Alternative Medicines. 2005;**2**(1):46-61. DOI: 10.4314/ajtcam.v2i1.31103

[130] Azzi L, Deluche C, Gévaudant F, Frangne N, Delmas F, Hernould M, et al. Fruit growth-related genes in tomato. Journal of Experimental Botany. 2015;**66**(4):1075-1086. DOI: 10.1093/jxb/eru527

[131] Basith S, Cui M, Hong S, Choi S. Harnessing the therapeutic potential of capsaicin and its analogues in pain and other diseases. Molecules. 2016;**21**(8):966. DOI: 10.3390/molecules21080966

[132] Maji AK, Banerji P. Phytochemistry and gastrointestinal benefits of the medicinal spice, *Capsicum annuum* L. (chilli): A review. Journal of Complementary and Integrative Medicine. 2016;**13**(2):97-122. DOI: 10.1515/jcim-2015-0037

[133] Gottardi D, Bukvicki D, Prasad S, Tyagi AK. Beneficial effects of spices in food preservation and safety. Frontiers in Microbiology. 2016;**7**(1394):1-20. DOI: 10.3389/fmicb.2016.01394

[134] Pérez-Gálvez A, Viera I, Roca M. Carotenoids and chlorophylls as antioxidants. Antioxidants. 2020;**9**(6):505. DOI: 10.3390/antiox9060505

[135] Howard LR, Talcott ST, Brenes CH, Villalon B. Changes in phytochemical and antioxidant activity of selected pepper cultivars (Capsicum species) As influenced by maturity. Journal of Agricultural and Food Chemistry. 2000;**48**(5):1713-1720. DOI: 10.1021/jf990916t

[136] Lim TK. Edible Medicinal and Non-medicinal Plants: Volume 10, Modified Stems, Roots, Bulbs. Berlin Heidelberg: Springer; 2016

[137] Kamiya K, Tanaka Y, Endang H, Umar M, Satake T. Chemical constituents of *Morinda citrifolia* fruits inhibit copper-induced low-density lipoprotein oxidation. Journal of Agricultural and Food Chemistry. 2004;**52**(19):5843-5848. DOI: 10.1021/jf040114k

[138] Clark R, Lee S-H. Anticancer properties of capsaicin against human cancer. Anticancer Research. 2016;**36**(3):837-843

[139] Liu N-C, Hsieh P-F, Hsieh M-K, Zeng Z-M, Cheng H-L, Liao J-W, et al. Capsaicin-mediated tNOX (ENOX2) up-regulation enhances cell proliferation and migration in vitro and in vivo. Journal of Agricultural and Food Chemistry. 2012;**60**(10):2758-2765. DOI: 10.1021/jf204869w

[140] Pabalan N, Jarjanazi H, Ozcelik H. The impact of capsaicin intake on risk of developing gastric cancers: A meta-analysis. Journal of Gastrointestinal Cancer. 2014;**45**(3):334-341. DOI: 10.1007/s12029-014-9610-2

[141] Sanati S, Razavi BM, hosseinzadeh, hossein. A review of the effects of Capsicum annuum L. and its constituent, capsaicin, in metabolic syndrome. Iranian Journal of Basic Medical Sciences. 2018;**21**(5):439-448. DOI: 10.22038/ijbms.2018.25200.6238

[142] Ramawat KG, Mérillon JM. editors. Natural Products. Berlin: Springer; 2013:1397-1437. DOI: 10.1007/978-3-642-22144-6_184

[143] Luo X-J, Peng J, Li Y-J. Recent advances in the study on capsaicinoids and capsinoids. European Journal of Pharmacology. 2011;**650**(1):1-7. DOI: 10.1016/j.ejphar.2010.09.074

[144] Srinivasan K. Biological activities of red pepper (*Capsicum annuum*) and its pungent principle capsaicin: A review. Critical Reviews in Food Science and Nutrition. 2016;**56**(9):1488-1500. DOI: 10.1080/10408398.2013.772090

[145] Derry S, Rice ASC, Cole P, Tan T, Moore RA. Topical capsaicin (high concentration) for chronic neuropathic pain in adults. In: The Cochrane Collaboration, editor. Cochrane Database of Systematic Reviews (p. CD007393.pub3). John Wiley & Sons, Ltd.; 2013. DOI: 10.1002/14651858.CD007393.pub3

[146] Huang W, Cheang WS, Wang X, Lei L, Liu Y, Ma KY, et al. Capsaicinoids but not their analogue Capsinoids lower plasma cholesterol and possess beneficial vascular activity. Journal of Agricultural and Food Chemistry. 2014;**62**(33):8415-8420. DOI: 10.1021/jf502888h

[147] Hsu Y-J, Huang W-C, Chiu C-C, Liu Y-L, Chiu W-C, Chiu C-H, et al. Capsaicin supplementation reduces physical fatigue and improves exercise performance in mice. Nutrients. 2016;**8**(10):648. DOI: 10.3390/nu8100648

[148] Ago H, Miyano M. Recent advances in biology of Cysteinyl leukotriene. Nihon Kessho Gakkaishi. 2010;**52**(1):69-75. DOI: 10.5940/jcrsj.52.69

Chapter 2

Padrón Peppers, Some Are Hot, and Some Are Not

José Díaz, Raquel Núñez-Fernández and Javier Veloso

Abstract

Some peppers are pungent due to the presence of their secondary metabolite contents—capsaicinoids. The ability to synthesize capsaicinoids is a genetic trait, but the control of their accumulation in the fruit is more complex than just biosynthesis. Besides biosynthesis, other metabolic pathways, such as oxidation and conjugation, are also involved in capsaicinoid homeostasis. Moreover, all these pathways are modulated by different factors, namely plant hormones, transcription factors, ontogeny, and the environment, including both abiotic and biotic agents. In the present chapter, the present knowledge about the control of capsaicin metabolism in pepper is reviewed. Based on the literature and our own experience, there is a correlation between pungency and lignification. We have a clue about the reason: capsaicinoid and lignin metabolic pathways are related, and their biosynthesis predate from the same intermediate compounds. Finally, this chapter mainly focuses on the cultivar Padrón, a pungent variety used in our experiments because of its economic and cultural value.

Keywords: *Capsicum*, capsaicinoids, lignin, pepper, pungency

1. Introduction

The title of this chapter is a translation of a local saying in Galicia (Northwest of Spain) that literally states in the Galician language, "Os pementos de Padrón, uns pican e outros non". The saying refers to the typical local dish consisting of green, immature fruits of pepper fried in oil and salted. When a dish of these peppers is served, you find that most fruits are sweet or mild, but sometimes you eat one that is extremely hot by chance. Padrón pepper production is of great economic relevance not only in Galicia but also in other parts of Spain, and even it is produced in other countries like Marocco (**Figure 1**).

These pepper fruits belong to a local landrace called "Padrón," which is characterized by the presence of capsaicinoids (the pungent substances in peppers), but in low amounts when fruits are immature [1]. The origin of the landrace can be traced back from the seeds that were taken in the 17th Century from Tabasco (Mexico) to the monastery of Herbón (a place in the municipality of Padrón) by monks of the Franciscan order [2]. The monks transferred seeds to the local farmers. The landrace arose from local breeding and established the tradition of restricting seed transfer within the local community. Interestingly, Padrón seeds are always part of women's dowry.

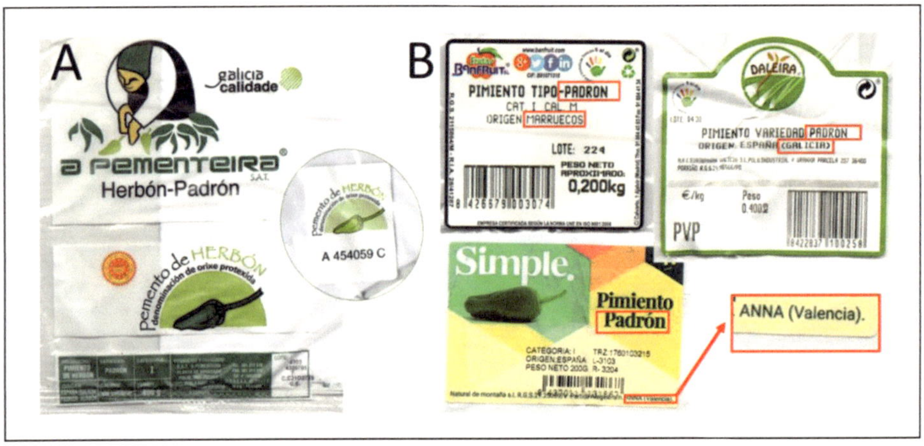

Figure 1.
A). Commercial bag of PDO "Pemento de Herbón" with the PDO label. B). Three different labels of Padrón commercial brands.

Figure 2.
Commercial products include Padrón peppers as an ingredient or as a motto. A) Paprika powder made with PDO Herbón peppers. B) PDO Herbón pepper jam. C) Hot candies. D) PDO Herbón chocolates. E) Padrón pepper sauce mimicking tabasco. F) Padrón pepper liquor. G) t-shirt with a "Galician roulette" pun regarding the Galician saying. H) Socks. I) Key chain.

The quality of the fruit, not only the heat but also other organoleptic traits, is considered of great importance by consumers. To guarantee this quality, some pepper growers in Padrón and other closer municipalities promoted the creation by the European Union of the Protected Designation of Origin (PDO) Pemento de Herbón, established in 2010.

On the other hand, the popularity of these peppers has led to the release of numerous commercial products, which include these peppers as an ingredient (paprika, jam, sauces, candies, chocolates, liquors, etc.) or items like t-shirts, socks, etc. as a part of Padrón-pepper passion (**Figure 2**).

Padron pepper consumers are divided into those who like such a pungency Russian roulette and those who prefer only sweet or hot fruits. Therefore, finding technical solutions to control the heat level is essential. Furthermore, it is not only a local issue: many other pepper landraces and cultivars are hot or sweet, and this trait has to be preserved when pepper lines are bred for, e.g., plant resistance to diseases, pests, and abiotic stress. Consumers expect a specific level of pungency (heat) in a particular cultivar's fruit, which has to be bear in mind to achieve success in sales. Moreover, even a pepper with the pungency (heat) trait can show a modulation of the level of capsaicinoids due to several factors [3].

This chapter reviews the present knowledge about pepper pungency and its control.

2. To be or not to be hot: capsaicinoid presence as a genetic trait

Some peppers are pungent, and others are not. Pepper pungency is a trait conferred by the presence of the functional allele of *Pun1* gene [4]. It is considered to be a dominant trait [5]. Still, the recent report of hot pepper hybrids derived from the crossing of two non-pungent parents [6] makes such a heterosis effect on capsaicinoid accumulation deserve further studies in the future.

The *Pun1* gene (also known as AT3) has been demonstrated to be involved in capsaicinoid biosynthesis. Firstly, there is a correlation between its expression with capsaicinoid accumulation and the lack of the pungent trait in pepper lines with mutations in this gene [4, 7]. Moreover, in the last years, several authors have demonstrated that virus silencing (VIGS) of the gene in pepper leads to a decrease in capsaicinoid accumulation [4, 8–10]. It should be noted that the pepper plant is recalcitrant to Agrobacterium transformation [3, 10]; therefore, the usual functional approach by genetic transformation of the plant has not been possible so far.

Another approach was to demonstrate that the protein encoded by Pun1/AT3 is the so-called capsaicin synthase. Ogawa et al. [9] expressed Pun1 protein in *Escherichia coli*, purified it, and obtained antibodies anti-Pun1, which inhibited capsaicinoid formation in an assay using pepper protoplasts. However, they did not assay the enzymatic activity of purified Pun1 itself. Recently, Muratovska et al. [11] expressed Pun1/AT3 and another aminotransferase gene (*CaAT*) in *Saccharomyces cerevisiae* and proved that both lines of transformed yeasts were able to synthesize the capsaicinoid nonivamide. Finally, Milde et al. [12] not only purified Pun1 expressed in *E. coli* but also demonstrated the ability of the purified protein to catalyze the capsaicin synthase reaction, that is, the transformation of trans 8-methyl-6-nonenoyl-CoA and vanilloylamine into capsaicin. Thus, Pun1 is a bonafide capsaicin synthase. However, the possibility that other pepper proteins, such as CaAT, could also be involved in the biocatalysis of this reaction cannot be excluded [11].

3. Being hot, but how much? Regulation of capsaicinoid levels

The presence of capsaicin synthase (Pun1/AT3 or other protein) is enough to allow the biosynthesis of capsaicinoids in a pepper line because their precursors are synthesized from common plant pathways. However, there are other pathways involved in capsaicin metabolism. Moreover, all those pathways are targets of fine-tune modulation in response to variations of different endogenous and exogenous factors. Hence, it is worthwhile to look at the present knowledge of regulation.

3.1 Capsaicinoid metabolic pathways

The capsaicin synthase reaction is the final step of the pathway, converting vanillylamine and 8-methyl-6-nonenoyl-Coa into capsaicin. The two substrates of the last reaction are synthesized in two different pathways. Thus, the aromatic/phenolic moiety comes from phenylalanine, and the aliphatic moiety from valine (**Figure 3**). More detailed information on this pathway can be found in [5, 13]. Branches of this pathway lead to the biosynthesis of capsinoids and capsaicinoids (details in [13]), non-pungent analogs of capsaicinoids that are usually synthesized in pepper lines with low content of the latter because of mutations in *Pun1*. Capsaicinoid biosynthesis shares precursors with other pathways. Thus, phenolic compounds are required to synthesize lignins, some phytoanticipins, flavonoids, etc. Therefore, there is competition among the different pathways that predate the same precursors.

Oxidases such as peroxidases and polyphenol oxidases can degrade capsaicinoids in pepper (**Figure 3**). Peroxidases can catalyze capsaicin oxidation *in vitro*, and their expression is correlated with capsaicin decrease at the end of fruit

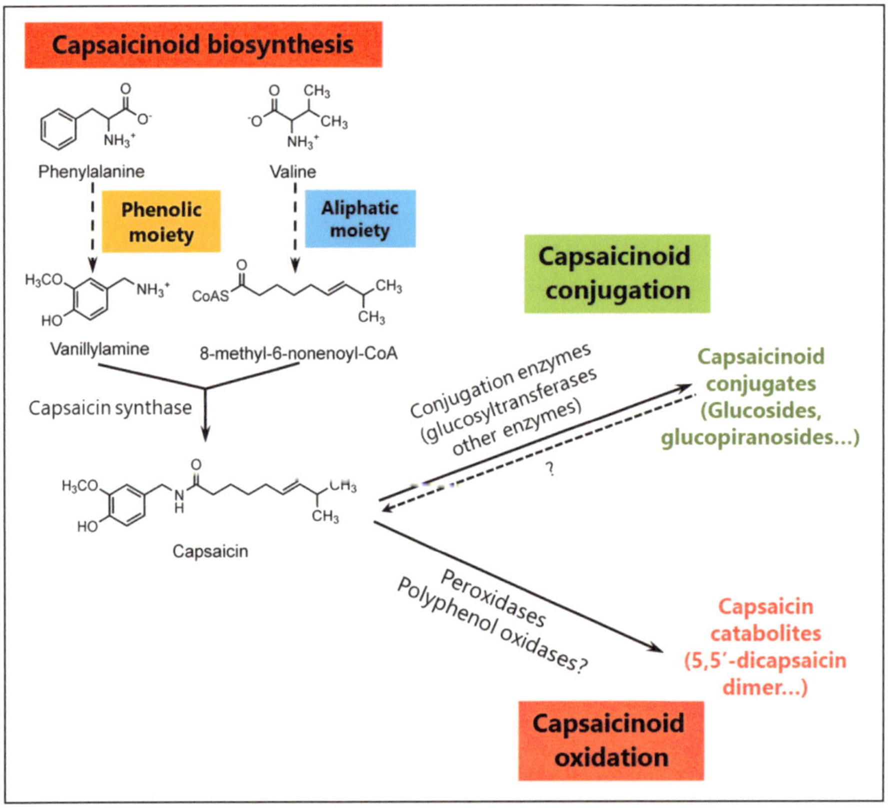

Figure 3.
A general overview of the pathways involved in capsaicinoid metabolism: biosynthesis, oxidation, and conjugation. Original figure created by the authors based on the cited literature.

development [5]. The main products of such oxidation are capsaicin dimers, such as 5,5′-dicapsaicin [5], which are present in natural sources such as pepper cell cultures and fruits [5, 14]. Immunoinhibition assays of capsaicin oxidation by peroxidases [14] supported the idea that other types of enzymes, such as polyphenol oxidases, could also be partially responsible for capsaicin oxidation [5].

Another metabolic fate of capsaicinoids is conjugation with other molecules (**Figure 3**). Thus, glucosides and glucopiranosides of capsaicin have been found in pepper fruits [5, 15]. However, we cannot exclude that other conjugated forms of capsaicinoids, e.g., with other sugars or amino acids, could be present in peppers.

3.2 Hormone and transcriptional regulation

3.2.1 Plant hormones

A previous review [5] summarized some evidence of plant hormone regulation of capsaicin metabolism, pointing to the role of ethylene, jasmonates, and salicylic acid, based on data from exogenous application of the hormones to cell cultures and plants. Since then, several pieces of information have been published, but the most interesting are those that have used new approaches. Thus, several transcription factors involved in capsaicin biosynthesis (see Section 3.2.2) are responsive to plant hormones such as ethylene (CcERF2, [16]) or jasmonates (CaMYB108, [17]). The evidence of the ethylene regulation of the capsaicinoid biosynthesis is clear: CcERF2 silencing resulted in decreased capsaicin accumulation, and the treatment of peppers with inhibitors of ethylene perception (1-methylcyclopropene) and biosynthesis (piperazine) also leads to a reduction of the pungent compounds in the fruit [16]. In the case of jasmonates, silencing a jasmonate-responsive transcription factor leads to a decrease in capsaicin and dihydrocapsaicin accumulation [17].

Moreover, the expression of a gene encoding an enzyme involved in jasmonate biosynthesis (2-oxophytoeienoic acid reductase) in pepper fruit is correlated with the stage of development when capsaicin accumulates [18]. However, more studies are needed before we can fully determine the hormone regulation of capsaicin pathways. For instance, limited studies have addressed the regulation of capsaicinoid oxidation or conjugation.

3.2.2 Transcriptional regulation

As in the case of hormones, gene silencing has been used in the last years to test the involvement of several transcription factors (TFs) in the biosynthesis of capsaicinoids. Most of them belong to the MYB type, but also, an AP2/ERF TF has been successfully proven to regulate the expression of capsaicin biosynthesis genes and capsaicinoid accumulation in the fruit (**Table 1**).

There are limited studies on the TF regulation of genes involved in capsaicinoid oxidation or conjugation. Other studies are based on the correlation between their expression and the expression of capsaicin biosynthesis genes or by computer-based analysis of their promoters. However, their involvement still has to be confirmed by gene silencing or other methods that provide equivalent information.

3.3 Organ and plant development

There are many reports of the trend of the capsaicinoid content in pepper fruit development: it increases continuously during most of the course of development,

Capsicum species	Gene	Type of TF	References
Capsicum chinense	CcMYB31*	MYB	[18]
Capsicum annuum	CaMYB31	MYB	[19]
C. annuum	CaMYB37	MYB	[20]
C. annuum	CaMYB48	MYB	[21]
C. annuum	CaMYB108	MYB	[17]
C. chinense	CcERF2	AP2 superfamily	[16]

**In this case, overexpression of the gene in pepper was also used.*

Table 1.
Transcription factors (TFs) proved to regulate capsaicin biosynthesis in pepper by gene silencing.

and just at the end of the process, it decreases [3]. In most cases, only the genes or proteins involved in the biosynthesis pathway were studied. Furthermore, usually, the study of competition among different capsaicinoid pathways was not addressed. However, Estrada et al. [1] demonstrated a negative correlation between lignin deposition and capsaicinoid accumulation, pointing to such competition (**Figure 4**).

Intriguingly, peroxidases and polyphenol oxidases (e.g., laccases) can affect capsaicinoid accumulation in at least two different ways: participating in the oxidation of these compounds or driving the flow of phenolic compounds to the lignin biosynthesis pathway by the catalysis of the last reaction of the pathway (**Figure 4**). In any case,

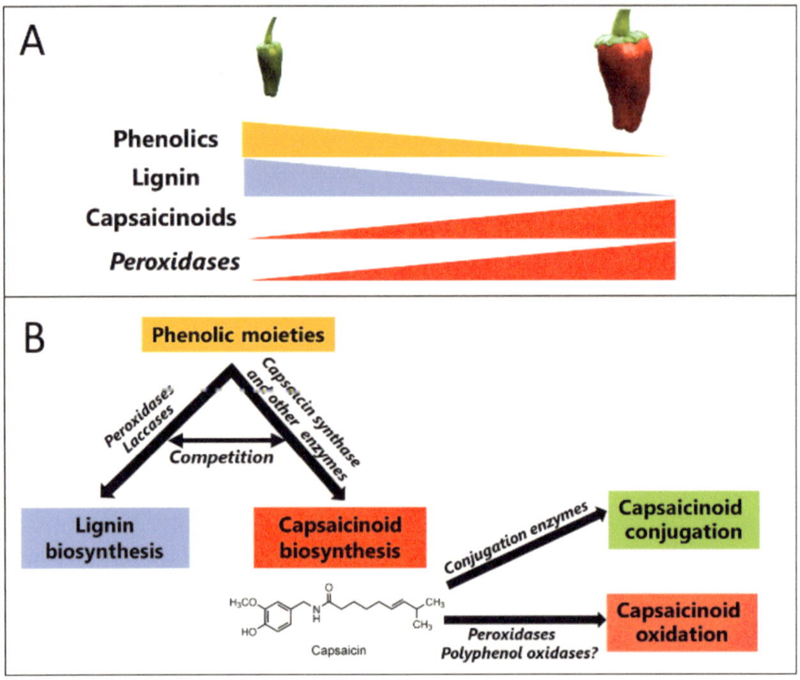

Figure 4.
Competition among different biosynthetic and catabolic pathways affects capsaicinoid levels in pepper fruit. A) Trends in the levels of capsaicinoids, phenolics, lignin, and peroxidases in Padrón pepper fruit (based on data from [1]). B) Overview of the involved pathways. Original figures created by the authors based on the cited literature.

this competition is underexplored, and conjugation has also been an oversight, even though it may be linked to the transport in vegetative organs.

Capsaicinoids have been detected in vegetative organs, but eliminating the floral buds and preventing fruit formation leads to the absence of these pungent compounds in leaves and stems [22]. This suggests that capsaicinoids could be transported from fruits to other organs of plants, but so far, this was not confirmed. Moreover, exogenous feeding of capsaicin to the roots of vegetative pepper plants does not lead to capsaicin presence in aerial organs [23]. Capsaicin is a compound not soluble in water, making its transport difficult into the plant. Maybe capsaicin conjugates, more soluble than capsaicin itself, are the compounds transported throughout the plant. To our knowledge, such a possibility has not been explored so far. Indeed, there is a lack of studies regarding capsaicin conjugates in pepper plants.

The age of the plant also determines the amount of capsaicin in the fruit, and older plants usually show more pungent fruits [5]. Probably as a consequence, the Padron pepper fruits are also hotter at the end of the season, in September–October (**Figure 5**).

3.4 Environmental factors and capsaicinoids

Several recent publications have reviewed the effects of several environmental abiotic factors as light, temperature, mineral nutrition, water, etc., on capsaicinoid accumulation, showing that, overall, stress causes an increase in these compounds (**Figure 6A**) [3, 24]. Therefore, we have focused on 1) the metabolic consequences of that effect regarding lignification and 2) biotic stress.

As we stated above, lignification competes with capsaicinoid biosynthesis during the development of the fruit. Therefore, an increase in capsaicinoid levels caused by stress should lead to a decrease in the deposition of lignin. This was exactly what we observed in previous studies regarding mineral nutrition and watering (**Figure 6B**) [25, 26].

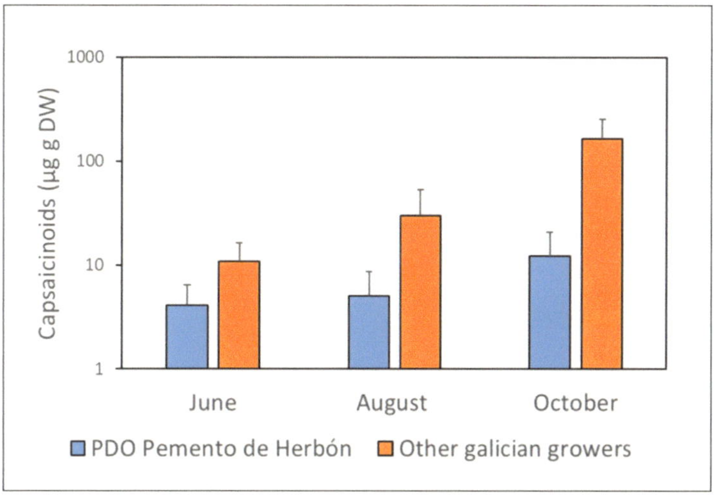

Figure 5.
Evolution of capsaicinoid levels in Padrón peppers during the commercial season. Samples were bought at Galician local markets during 2021 and 2022 (data from the Ph.D. thesis of Raquel Núñez-Fernández, in preparation).

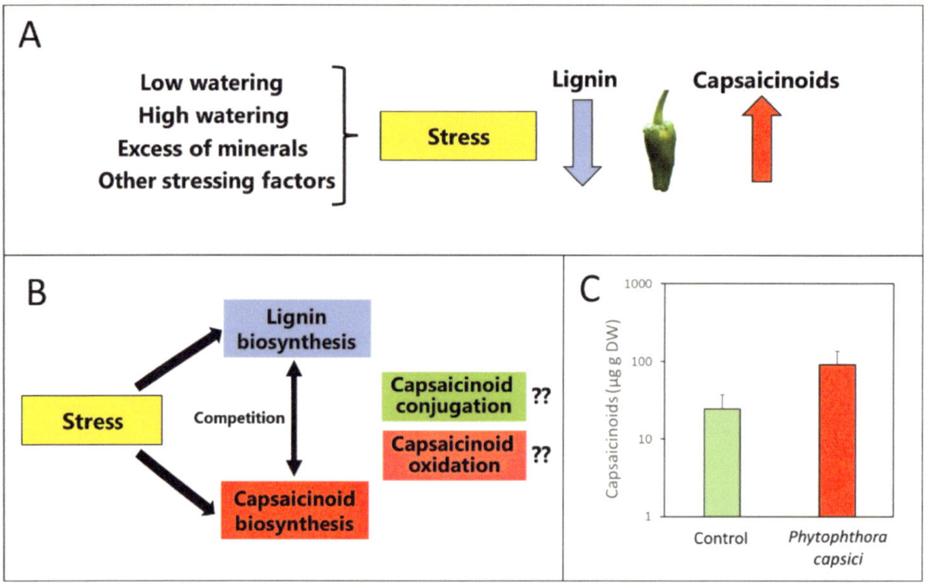

Figure 6.
Effects of stress on capsaicinoid levels. A) Stress causes an increase in capsaicinoids and a decrease in lignin in the fruit (based on data from [25, 26]). B) the stress-induced accumulation of capsaicinoids causes a reduction in the phenolic moieties that otherwise would be used in lignification. C) Phytophthora capsici infection causes stress in pepper plants, thus leading to increased capsaicinoid accumulation in the fruit (data from Ph.D. thesis of Raquel Núñez-Fernández, in preparation).

Biotic stress also can modulate the amount of capsaicinoids in the fruit. However, capsaicin quantification is usually oversight in agronomic trials testing biofertilizers, biostimulants, and biological control agents (BCAs), as well as studies where a pathogen or pest is used as a challenger. However, this analysis is worthwhile because the market expects a stable pungency level [3]. Thus, Saxena et al. [27] found that pepper plants treated with Trichoderma isolates used as BCAs caused an increase in capsaicin in the fruits of plants after *Colletotrichum truncatum* challenge. Khan et al. [28] tested the endophyte *Penicillium resedanum* as a potential tool to alleviate drought stress in pepper. They reported an increase in the capsaicin levels in the fruits of the plants treated with this fungal endophyte. In our experiments, we observed that the stress in plants inoculated with *Phytophthora capsici* leads to increased capsaicin in the fruit **Figure 6C**). Therefore, in the last years, we included the analysis of capsaicinoids in greenhouse trials while testing BCAs and resistance inducers.

4. Conclusions and perspectives

Control of the capsaicin levels beyond a simple absence/presence trait is necessary. Producing pepper fruits with uniform and predictable levels of pungency is a challenge that needs further insight into the regulation of both the biosynthesis and the catabolic pathways of capsaicinoids. The knowledge of oxidation and conjugation pathways has not advanced much in the last two decades, probably because scientists paid little attention to them. On the other hand, genomic, transcriptomic, proteomic, and metabolomics studies have flourished in the last decade. Still, more molecular and physiological studies are required, particularly studies proving the function of the genes by silenced lines or mutants if the recalcitrant nature of pepper for genetic transformation is overcome. There is room for improvement in our knowledge of these physiological processes, and basic science advances will lead to applied advances. Pungent compounds can be used not only as food but also can be useful in medicine as drugs or in agriculture as fungicides or pesticides [3, 23], and all these applications need a guarantee of stable supply.

Acknowledgements

This research was funded by grant RTA2017-00019-C03-03 from Ministerio de Ciencia e Innovación.

Conflict of interest

The authors declare no conflict of interest.

Author details

José Díaz[1]*, Raquel Núñez-Fernández[1] and Javier Veloso[1,2]

1 Faculty of Sciences, FISAPLANT Research Group, Department of Biology, Universidade da Coruña, A Coruña, Spain

2 Department of Functional Biology, Higher Polytechnic Engineering School, Universidade de Santiago de Compostela, Lugo, Spain

*Address all correspondence to: josefv@udc.es

References

[1] Estrada B, Bernal MA, Díaz J, Pomar F, Merino F. Fruit development in *Capsicum annuum*: Changes in capsaicin, lignin, free phenolics and peroxidase patterns. Journal of Agricultural and Food Chemistry. 2000;**48**:6234-6239. DOI: 10.1021/jf000190x

[2] Official Journal of the European Union. Publication of an application pursuant to Article 6(2) of Council Regulation (EC) No 510/2006 on the protection of geographical indications and designations of origin for agricultural products and foodstuffs. COUNCIL REGULATION (EC) No 510/2006 'PEMENTO DE HERBÓN' EC No: ES-PDO-0005-0509-15.11.2005. [Internet]. 2009. Available from: https://eur-lex.europa.eu/legal-content/EN/TXT/PDF/?uri=CELEX:52009XC1218(07)&from=EN. [Accessed January 3, 2023]

[3] Naves ER, de Ávila SL, Sulpice R, Araújo WL, Nunes-Nesi A, Peres LEP, et al. Capsaicinoids: Pungency beyond Capsicum. Trends in Plant Science. 2019;**24**:109-120. DOI: 10.1016/j.tplants.2018.11.001

[4] Stewart C Jr, Kang B-C, Liu K, Mazourek M, Moore SL, Yoo EY, et al. The Pun1 gene for pungency in pepper encodes a putative acyltransferase. The Plant Journal. 2005;**42**:675-688. DOI: 10.1111/j.1365-313X.2005.02410.x

[5] Díaz J, Pomar J, Bernal MA, Merino F. Peroxidases and the metabolism of capsaicin in *Capsicum annuum* L. Phytochemistry Reviews. 2004;**3**:141-157. DOI: 10.1023/B:PHYT.0000047801.41574.6e

[6] Naves ER, Scossa F, Araújo WL, et al. Heterosis for capsaicinoids accumulation in chili pepper hybrids is dependent on parent-of-origin effect. Scientific Reports. 2022;**12**:14450. DOI: 10.1038/s41598-022-18711-w

[7] Stewart C Jr, Mazourek M, Stellari GM, O'Connell M, Jahn MM. Genetic control of pungency in *C. chinense* via the *Pun1* locus. Journal of Experimental Botany. 2007;**58**:979-991. DOI: 10.1093/jxb/erl243

[8] Arce-Rodríguez ML, Ochoa-Alejo N. Silencing AT3 gene reduces the expression of pAmt, BCAT, kas, and Acl genes involved in capsaicinoid biosynthesis in chili pepper fruits. Biologia Plantarum. 2015;**59**:477-484. DOI: 10.1007/s10535-015-0525-y

[9] Ogawa K, Murota K, Shimura H, et al. Evidence of capsaicin synthase activity of the Pun1-encoded protein and its role as a determinant of capsaicinoid accumulation in pepper. BMC Plant Biology. 2015;**15**:93. DOI: 10.1186/s12870-015-0476-7

[10] Kim J, Park M, Jeong ES, et al. Harnessing anthocyanin-rich fruit: A visible reporter for tracing virus-induced gene silencing in pepper fruit. Plant Methods. 2017;**13**:3. DOI: 10.1186/s13007-016-0151-5

[11] Muratovska N, Grey C, Carlquist M. Engineering *Saccharomyces cerevisiae* for production of the capsaicinoid nonivamide. Microbial Cell Factories. 2022;**21**:106. DOI: 10.1186/s12934-022-01831-3

[12] Milde R, Schnabel A, Ditfe T, Hoehenwarter W, Proksch C, Westermann B, et al. Chemical synthesis of trans 8-methyl-6-nonenoyl-CoA and functional expression unravel capsaicin synthase activity encoded by the Pun1

locus. Molecules. 2022;**27**:6878. DOI: 10.3390/molecules27206878

[13] García T, Veloso J, Díaz J. Properties of vanillyl nonanoate for protection of pepper plants against *Phytophthora capsici* and *Botrytis cinerea*. European Journal of Plant Pathology. 2018;**150**:1091-1101. DOI: 10.1007/s10658-017-1352-0

[14] Zamudio-Moreno E, Echevarría-Machado I, Medina-Lara MF, Calva-Calva G, Miranda-Ham ML, Martínez-Estévez M. Role of peroxidases in capsaicinoids degradation in habanero pepper (*Capsicum chinense* Jacq.) plants grown under water deficit conditions. Australian Journal of Crop Science. 2014;**8**:448-454

[15] Elkhedir A, Iqbal A, Albahi A, et al. Capsaicinoid-glucosides of fresh hot pepper promotes stress resistance and longevity in *Caenorhabditis elegans*. Plant Foods for Human Nutrition. 2022;**77**:30-36. DOI: 10.1007/s11130-021-00939-y

[16] Wen J, Lv J, Zhao K, Zhang X, Li Z, Zhang H, et al. Ethylene-inducible AP2/ERF transcription factor involved in the capsaicinoid biosynthesis in Capsicum. Frontiers in Plant Science. 2022;**13**:832669. DOI: 10.3389/fpls.2022.832669

[17] Sun B, Zhu Z, Chen C, Chen G, Cao B, Chen C, et al. Jasmonate-inducible R2R3-MYB transcription factor regulates capsaicinoid biosynthesis and stamen development in Capsicum. Journal of Agricultural and Food Chemistry. 2019;**67**:10891-10903. DOI: 10.1021/acs.jafc.9b04978

[18] Zhu Z, Sun B, Cai W, Zhou X, Mao Y, Chen C, et al. Natural variations in the MYB transcription factor MYB31 determine the evolution of extremely pungent peppers. New Phytologist. 2019;**223**:922-938. DOI: 10.1111/nph.15853

[19] Arce-Rodríguez ML, Ochoa-Alejo N. An R2R3-MYB transcription factor regulates Capsaicinoid biosynthesis. Plant Physiology. 2017;**174**:1359-1370. DOI: 10.1104/pp.17.00506

[20] Liu Y, Zhang Z, Fang K, Shan Q, He L, Dai X, et al. Genome-wide analysis of the MYB-related transcription factor family in pepper and functional studies of CaMYB37 involvement in capsaicin biosynthesis. International Journal of Molecular Sciences. 2022;**23**:11667. DOI: 10.3390/ijms231911667

[21] Sun B, Zhou X, Chen C, Chen C, Chen K, Chen M, et al. Coexpression network analysis reveals an MYB transcriptional activator involved in capsaicinoid biosynthesis in hot peppers. Horticultural Research. 2020;**7**:162. DOI: 10.1038/s41438-020-00381-2

[22] Estrada B, Bernal MA, Díaz J, Pomar F, Merino F. Capsaicinoids in vegetative organs of Capsicum annuum L. in relation to fruiting. Journal of Agricultural and Food Chemistry. 2002;**50**:1188-1191. DOI: 10.1021/jf011270j

[23] Veloso J, Prego C, Varela MM, Carballeira R, Bernal A, Merino F, et al. Properties of capsaicinoids for the control of fungi and oomycetes pathogenic to pepper. Plant Biology (Stuttgart, Germany). 2014;**16**:177-185. DOI: 10.1111/j.1438-8677.2012.00717.x

[24] Uarrota VG, Maraschin M, de Bairros ÂFM, Pedreschi R. Factors affecting the capsaicinoid profile of hot peppers and biological activity of their non-pungent analogs (Capsinoids) present in sweet peppers. Critical

Reviews in Food Science and Nutrition. 2021;**61**:649-665. DOI: 10.1080/10408398.2020.1743642

[25] Estrada B, Pomar F, Díaz J, Merino F, Bernal MA. Effects of mineral fertilizer supplementation on fruit development and pungency in 'Padrón' peppers. The Journal of Horticultural Science and Biotechnology. 1999;**73**:493-497. DOI: 10.1080/14620316.1998.11511004

[26] Estrada B, Pomar F, Dıiaz J, Merino F, Bernal MA. Pungency level in fruits of the Padrón pepper with different water supply. Scientia Horticulturae. 1999;**81**:385-396. DOI: 10.1016/S0304-4238(99)00029-1

[27] Saxena A, Mishra S, Ray S, et al. Differential reprogramming of Defense network in *Capsicum annum* L. plants against *Colletotrichum truncatum* infection by phyllospheric and rhizospheric *Trichoderma* strains. Journal of Plant Growth Regulation. 2020;**39**:751-763. DOI: 10.1007/s00344-019-10017-y

[28] Khan AL, Shin J-H, Jung H-Y, Lee I-J. Regulations of capsaicin synthesis in *Capsicum annuum* L. by *penicillium resedanum* LK6 during drought conditions. Scientia Horticulturae. 2014;**175**:167-173. DOI: 10.1016/j.scienta.2014.06.008

Chapter 3

Pharmacological Properties and Health Benefits of *Capsicum* Species: A Comprehensive Review

Kalaiyarasi Dhamodharan, Manobharathi Vengaimaran and Mirunalini Sankaran

Abstract

Since the start of history, natural medicine has been of great interest and attention to humankind. A heap of empirical research indicates that spices have undoubtedly made our lives more interesting and may also make them longer. *Capsicum* is a highly regarded indispensable spice all over the globe for its umpteen culinary and medicinal facets. It has been used for more than 7000 years in Mexico and is believed to have originated in tropical Central America. Mainly, this botanical contains a good source of vitamin C, vitamin A, vitamin E, vitamin B5, potassium, magnesium, iron, calcium, phosphorus, and carotenoids. Interestingly, *capsicum* phenolic compounds are helpful in preventing and treating many ailments. So, it intends as a beneficial milestone in the pharmaceutical industry and a boon to humanity. This chapter highlights the tremendous pharmacological uses and health benefits of *capsicum* species and its active compounds in multifarious aspects.

Keywords: *Capsicum*, spice, phenolic compounds, nutrition, medicine

1. Introduction

Capsicum is a popular vegetable and spice crop belonging to the nightshade family Solanaceae, which is amply cultivated for its succulent berries and seeds in tropical and subtropical climate regions all around the globe. The word "*Capsicum*" comes from the Greek word kapsimo, meaning "to bite" or "to swallow." Astonishingly, *Capsicum* pods have been well known since the beginning of civilization in the Western hemisphere and have been part of the human diet since 7500 BC. Habitually, people eat this botanical spice in raw, dried, and cooked form, and it is also used in making paste, pickle, and sauce. Although, from place to place, the name and type of *Capsicum* berries vary, the most common variety is called "pepper or chili pepper," which itself can vary greatly in color, shape, size, appearance, flavor, and pungency. Basically, the color diversity of *Capsicum* fruit is linked to the presence of pigments like chlorophyll (green), anthocyanins (violet/purple), α-carotene, β-carotene, zeaxanthin, lutein, and β-cryptoxanthin (yellow/orange) [1]. Surprisingly, approximately 35 species of *capsicum* exist in nature; only five have been domesticated for human consumption, namely *Capsicum annuum* (ancho/poblano, bell, cayenne, thai, jalapeno, paprika, pimiento, piquin, and serrano), *Capsicum baccatum* (aji amarillo, aji limon, criolla sella, malawi piquante, and bishop's crown), *Capsicum chinense* (scotch bonnets, trinidad scorpions, bhut jolokia, and carolina reaper), *Capsicum frutescens* (tabasco, bird-eye, kambuzi, malagueta, and siling labuyo), and *Capsicum pubescens* (rocoto pepper) (**Figure 1**) [2]. Of these species, the *Capsicum annuum* is the most economically important crop due to its pungent odor and taste.

Chili peppers are perennial woody plants grown as herbaceous annuals. It is said to be the first-ever domesticated crop in America [3]. The size of the plant can range from two to four feet tall, depending on the species. Typically, leaves are smooth, simple, entire, glabrous, and flat. The flowers are usually solitary, creamy white, and the seeds are straw-colored. **Figure 2** highlights the different types of chili plants. In addition, chili plants are grown for ornamental purposes, owing to their bright, shining fruits with a diverse range of colors [4]. Most abundantly, chili crops are grown in Pakistan, India, China, Ethiopia, Myanmar, Mexico, Vietnam, Turkey, Peru, Ghana, Bangladesh, Japan, Africa, and America (**Table 1**). As per 2019 world production statistics, the total global produce of chili pepper is 38 million tons [5]. China ranks first, producing over 18,978,027 tons of chili in 2019. In terms of nutritional standpoint, chili is considered to be one of the most nutritionally dense foods on earth, and it plays a vital role in alleviating human micronutrient deficiencies [6, 7]. Traditionalistically, it is harnessed in different systems of medicine to combat a wide variety of diseases and/or disorders due to the presence of therapeutically significant active constituents [8]. A total of 200 phytoconstituents have been identified from chilies [9]. Chili pepper's extremely hot or burning sensation is due to capsaicinoids, a family of compounds consisting of acid amides of vanillylamine and a C_8–C_{13}

Figure 1.
Five major domesticated species of Capsicum.

Figure 2.
Different types of chili plants.

Rank	Country	Production (tons)
1	China	18,978,027
2	Mexico	3,238,245
3	Turkey	2,625,669
4	Indonesia	2,588,633
5	Spain	1,402,380
6	Egypt	764.292
7	Nigeria	753.116
8	Algeria	675.168
9	United States of America	624.982
10	Tunisia	443.632

Source: FAOSTAT [5].

Table 1.
List of the top ten chili pepper producing countries in 2019.

branched-chain fatty acid [10]. Capsaicin and dihydrocapsaicin are the two prominent capsaicinoids present in chili peppers, accounting for over 90% of the total capsaicinoids [11]. Particularly, capsaicin has been at the center of intense research to elucidate the basis of its pharmacological properties and exploit its therapeutic potential [12, 13]. In recent times, this chemical substance has been employed as an analgesic in topical ointments, nasal sprays, and dermal patches to treat pain, typically in concentrations between 0.025 and 0.1%. It is also used to reduce the symptoms of peripheral neuropathy, such as postherpetic neuralgia caused by shingles [14]. Other capsaicinoids, such as nordihydrocapsaicin, homocapsaicin, and homodihydrocapsaicin, are present in small amounts in chili peppers, accounting for less than 10% of the total capsaicinoids [15]. Therefore, this chapter aims to discuss the nutritional value, phytochemical profile, pharmacological properties, and health benefits of *Capsicum* species.

2. Nutritional value and phytochemical constituents

Chili peppers are a good source of dietary fiber, riboflavin, thiamin, folate, niacin, iron, protein, phosphorus, and copper. Aside from that, it also contains high amounts of vitamin A, vitamin C, vitamin K, vitamin E, vitamin B6, potassium, and manganese [16]. The nutritional composition of chili peppers per 100 g is listed in **Table 2**. Chili fruits are also rich in many phytochemicals such as carotenoids (lutein, β-carotene,

S. No.	Types of nutrient	Amount
1	Water	88.02 g
2	Energy	40 kcal
3	Protein	1.87 g
4	Fat	0.44 g
5	Carbohydrate	8.81 g
6	Fiber	1.5 g
7	Sugars	5.3 g
	Minerals	
8	Calcium	14 mg
9	Iron	1.03 mg
10	Magnesium	23 mg
11	Phosphorus	43 mg
12	Potassium	322 mg
13	Sodium	9 mg
14	Zinc	0.26 mg
	Vitamins	
15	Vitamin C	143.7 mg
16	Thiamin	0.072 mg
17	Riboflavin	0.086 mg
18	Niacin	1.244 mg
19	Vitamin B-6	0.506 mg
20	Folate	23 μg
21	Vitamin A	48 μg
22	Vitamin E	0.69 mg

S. No.	Types of nutrient	Amount
23	Vitamin K	14 µg
	Lipids	
24	Fatty acids, total saturated	0.042 g
25	Fatty acids, total monounsaturated	0.024 g
26	Fatty acids, total polyunsaturated	0.239 g

Table 2.
Nutritional composition of chili pepper (per 100 g).

β-cryptoxanthin, zeaxanthin, violaxanthin, and capsanthin), capsaicinoids (capsaicin, dihydrocapsaicin, nordihydrocapsaicin, homocapsaicin, homodihydrocapsaicin, and nonivamide), and flavonoids (quercetin, luteolin, kaempferol, catechin, epicatechin, rutin, apigenin, myricetin, and cyanidin) [17, 18]. The following **Figure 3** shows the chemical structures of the various phytochemical constituents.

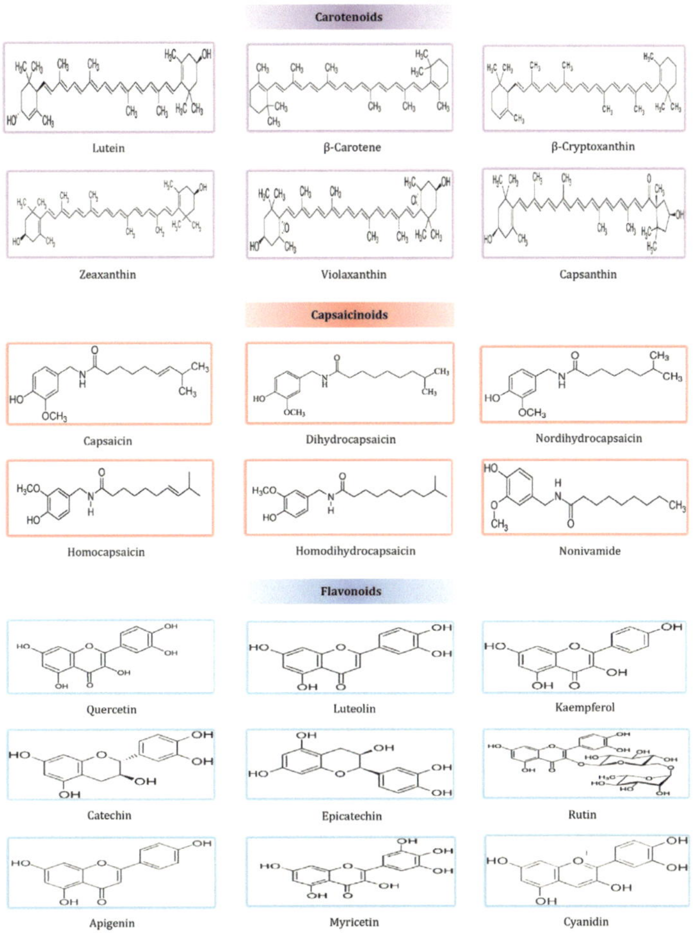

Figure 3.
Chemical structures of various phytochemical constituents in chili pepper.

3. Pharmacological uses

The mechanism behind the therapeutic potential of chili pepper has been apprised in several hefty pieces of literature. Chili is effective against a great number of ailments such as cancer, rheumatoid arthritis, bronchitis, macular degeneration, anemia, osteoporosis, coronary heart disease, diabetes, obesity, hypertension, sinus infection, migraine, neurological disorders, menopausal problems, and digestive complications [19–25]. **Figure 4** displays the pharmacological activities of chili pepper. Herein, the immense potential of chili in battling severe illnesses, as well as the mechanisms associated with health-promoting actions, have been illustrated in detail.

3.1 Anticancer activity

Cancer is a group of diseases characterized by uncontrolled growth and the spread of abnormal cells. It is the world's second leading cause of death, with a 10 million fatality rate annually [26]. A series of changes in the activities of cell cycle regulators are usually hooked up for cancer development and progression [27]. Generally, cancers are often named for the organ or cell type where the abnormal cells first form. Lung, prostate, colorectal, stomach, and liver cancer are the most prevalent types of cancer in men, while breast, colorectal, lung, cervical, and thyroid cancer are the

Figure 4.
Pharmacological activities of chili pepper.

most endemic among women [26]. Current chemotherapeutic drugs are enormously utilized to destroy cancer cells. Still, in addition to targeting the diseased cells, it also kills healthy blood cells, skin, stomach, hair follicles, bone marrow, etc. As a

Cancer type	Dose or concentration/ duration of application/ ingestion	Effect/mechanism	Experimental model	References
Pancreatic cancer	200 μM/L for 24 h	Endoplasmic reticulum stress (ERS) mediated apoptosis	*In vitro* (PANC-1)	[38]
	150 μM/L for 24 h		*In vitro* (SW1990)	
	5 mg/kg oral administration for 35 days	Suppresses tumor growth by inhibiting Trx and activating ASK1	*In vivo* (athymic nude mice)	[39]
Colon cancer	200 and 300 μM for 24 h	Peroxisome proliferator-activated receptor γ (PAPR γ) mediated apoptotic cell death	*In vitro* (HT-29)	[40]
	10–400 μM for 24 h	G0/G1 cell cycle arrest and induce apoptosis	*In vitro* (HCT116 and LoVo)	[41]
	1 and 3 mg/kg intraperitoneal administration for 30 days	Inhibition of tumor growth	*In vivo* (BALB/c athymic nude mice)	[42]
Liver cancer	50 μM for 36 h	Inhibition of migration and invasion	*In vitro* (SMMC-7721 and HepG2)	[43]
	50 μM for 3 h	Inhibition of adhesion		
Lung cancer	50 μM for 72 h	Inhibition of cell proliferation through E2F pathway	*In vitro* (H69, H82, DMS53, and DMS114)	[44]
	10 mg/kg in an AIN-76A based diet		*In vivo* (nude mice)	
Prostate cancer	1–100 μM for 24–72 h	Inhibition of PI3K/Akt/mTOR axe and modulates autophagy	*In vitro* (PC-3 and LNCaP)	[45]
	5 mg/kg, 3 days a week, oral administration for 30 weeks	Inhibition of migration, invasive, neuroendocrine differentiation process, and upregulation of the tumor suppressor protein $p27^{Kip1}$	*In vivo* (TRAMP mice)	[46]
Breast cancer	200 μM for 72 h	Inhibition of cell growth via apoptosis and cell cycle arrest in the S phase	*In vitro* (MCF-7 and BT-20)	[47]
	10 mg/kg, once in 3 days, intraperitoneal administration for 21 days	Inhibition of proliferation and induces apoptosis via down-regulating FBI-1 mediated NF-κB pathways (Ki-67, Bcl-2, Bax, caspase 3, and survivin proteins)	*In vivo* (BALB/c athymic nude mice)	[48]
Bladder cancer	50–300 μM for 48 h	Suppresses cell proliferation, induces cell cycle arrest and ROS production through modulating FOXO3a-mediated pathways	*In vitro* (5637 and T24)	[49]
	20 mg/kg, every two days for four weeks, injected peritumoral area		*In vivo* (NOD/ SCID mouse)	

Table 3.
Anticancer effects of capsaicin on various cancers in in vitro and in vivo models.

consequence of the undesirable properties and side effects of synthetic drugs, natural products have become increasingly popular over the past few decades. Capsaicin, the spicy ingredient of hot chili peppers, exhibits anti-neoplastic activity in a vast number of cancers like pancreatic cancer, colon cancer, liver cancer, lung cancer, prostate cancer, breast cancer, bladder cancer, and skin cancer [28–36]. The significant anticancer capacity of capsaicin targets multiple signaling pathways and cancer-associated genes in different phases of tumor development, including initiation, promotion, progression, and metastasis [37]. **Table 3** shows that various *in vitro* and *in vivo* models have been used to demonstrate the anticancer effects of capsaicin.

3.1.1 Pancreatic cancer

Pancreatic cancer, one of the most lethal malignancies, is the seventh leading cause of cancer-related fatality globally. This disorder is broken down into two forms: pancreatic adenocarcinoma (85%, with a very poor prognosis) and pancreatic neuroendocrine tumors [50]. In patients with advanced pancreatic cancer, the survival rate is less than one year. Numerous studies have explored the possibility of improving survival in pancreatic cancer with new therapies. Over the past few years, researchers have studied the effects of capsaicin on various pancreatic cancer cell lines, including BxPC-3, AsPC-1, PANC-1, SW1990, MiaPaCa-2, and L3.6pl. Based on the results of these studies, anti-proliferative activities of capsaicin are mainly attributed to the inhibition of oxidative stress and angiogenesis, cell cycle regulation, and apoptosis induction [38, 51, 52]. The first report on the involvement of endoplasmic reticulum stress (ERS) in the induction of apoptosis in PANC-1 and SW1990 cells using capsaicin was described by Lin et al. [38]. The authors demonstrated the potency of capsaicin on the mRNA expression of two key ERS markers (GRP78 and GADD153) in PANC-1 and SW1990 cells. According to real-time PCR analysis, capsaicin significantly increased the mRNA expression of GRP78 and GADD153 in a time and dose-dependent manner, suggesting ERS-mediated apoptosis and cell growth inhibition.

3.1.2 Colon cancer

Globally, colon cancer is one of the most prevalent forms of cancer, posing a major public health threat. In 2020, approximately 1,148,515 people were diagnosed with colon cancer, and 576,858 people died from this disease worldwide [53]. The onset of colon cancer is associated with excessive cell proliferation and dysregulation of both cell-cycle progression and apoptosis. Additionally, "neoangiogenesis" plays an essential role in the development, growth, and metastasis of colon tumors [54]. In the majority of cases, colon tumors are only diagnosed in the later stages of the disease, because they do not manifest as pain-like symptoms. Over time, several strides have been made in researching and treating colon cancer. However, its survival rate has not significantly improved. The five-year survival rate is still less than 15% due to the available therapeutic agents showing strong adverse effects and poor effectiveness [55]. Recent research reported that capsaicin has cytotoxic effects on different human colorectal cancer cell lines, including colo 205 and RKO [42, 56]. In a 2004 study, Kim et al. presented salient findings regarding the capsaicin-induced apoptotic cell death by activating peroxisome proliferator-activated receptor γ (PAPR γ) in HT-29 human colon cancer cells [40]. In addition, the latest study has shown that capsaicin mediates cell cycle arrest and apoptosis in two different human colon cancer cells (HCT116 and LoVo) via stabilizing and activating p53 in a time-dependent manner [41].

3.1.3 Liver cancer

Hepatocellular carcinoma is the most-encountered primary liver cancer in adults. Overall, the incidence rate of liver cancer is approximately four times higher in males than in females, and its pathogenesis is usually considered as an overlap of long-lasting processes, such as hepatic cytolysis, inflammation, liver regeneration, and fibrosis [57]. In hepatocellular carcinoma cell line HepG2, capsaicin-induced apoptosis with the involvement of intracellular Ca^{2+}, ROS, Bcl-2 family, cytochrome c protein expression, and caspase-3 activity [58]. Co-treatment with capsaicin and sorafenib potentially inhibits cell proliferation by activating caspase-9, PARP, AMPK, acetyl CoA carboxylase phosphorylation in HepG2 and Huh-7 cells [59].

3.1.4 Lung cancer

Lung cancer is the leading cause of mortality in both men and women, and it causes 1.8 million deaths annually. On the medical front, the prognosis of lung cancer is poor because it cannot produce noticeable signs and symptoms in the early stages. Being exposed to cigarette smoke/smoking is considered the most important factor involved in lung cancer development. Besides, environmental pollution and epigenetic alterations can also lead to lung cancer progression. This kind of cancer is broadly classified into two types: small cell lung cancer (SCLC) and non-small-cell lung cancer (NSCLC). NSCLC is the predominant type of lung cancer, accounting for about 85% of cases, while SCLC is responsible for 15% of lung cancer cases [60]. Capsaicin exhibited its therapeutic efficiency in lung cancer treatment by means of inhibiting Hypoxia-inducible factor (HIF)-1α accumulation by suppressing mitochondrial respiration in human lung cancer cells (H1299, H23, A549, and H2009) [61]. Furthermore, the time-dependent antitumor effects of capsaicin on lung cancer were also described in an *in vivo* and *in vitro* study on SCLC cells and nude mice. The study showed that capsaicin-induced apoptosis is mediated by transient receptor potential vanilloid subfamily member 6 (TRPV6), intracellular calcium, and calpain pathway [62]. TRPV6 is one of the most calcium selective ion channels in the TRPV family. Its main function is to regulate calcium transport, reabsorption, and homeostasis in epithelial tissues. Calpains are a family of calcium-dependent intracellular cysteine proteases that regulate multiple cellular processes.

3.1.5 Prostate cancer

Prostate cancer is the most common invasive malignancy among males. The incidence rate has increased in recent years in most regions of the world, perhaps due to improved detection methods with prostate-specific antigen (PSA) testing; however, the mortality rate has remained constant since the early 1900s. Androgen and androgen receptor (AR) play a critical role in the growth and maintenance of the prostate gland and the development of prostate tumors [63]. Prostate cancer may be connected with debilitating disease-related complications in an advanced stage, including painful bone metastases and urinary tract obstruction. microRNAs (miRNAs) are a class of small non-coding RNAs (ncRNAs) that regulate gene expression by repressing translation and have been proven to be implicated in the regulation of crucial processes, such as proliferation, differentiation, and apoptosis in various kinds of cancer [64]. Among the miRNAs, miR-449a functions as an important tumor suppressor in many types of tumors by targeting different genes. Recently, Zheng et al.

found that capsaicin inhibits the proliferation of AR-positive prostate cancer cells (C4-2 and LNCaP) by inducing the restoration of miR-449a [65]. Additional convergent pieces of evidence have shown that the capsaicin combined with brassinin and docetaxel synergistically kills human prostate cancer cells (PC-3 and LNCaP) through metabolic regulator AMP-activated kinase and apoptosis [66, 67].

3.1.6 Breast cancer

Breast cancer is the second most prevalent cancer worldwide and causes a high number of deaths among women every year. In the proliferation of breast cancer cells, NF-κB—the proinflammatory transcription factor plays a key role. It regulates more than 500 different genes and governs the expression of proteins engaged in cellular signaling pathways, leading to the development of malignancies and inflammation. Capsaicin displayed the ability to affect breast cancer cell proliferation by downregulating the FBI-1-Mediated NF-κB pathway [48]. Another target that acts on the proliferation of breast cancer cells is the human epidermal growth factor receptor-2 (HER-2), a tyrosine kinase (TK) receptor belonging to the EGFR family. A recent study by Thoennissen et al. showed capsaicin causes cell-cycle arrest and apoptosis in breast cancer cells (MCF-7, T47D, BT-474, SKBR-3, and MDA-MB231) via modulating the EGFR/HER-2 pathway [68].

Cyclin-dependent kinases (CDKs), a member of the serine/threonine-protein kinase family, can coordinate critical regulatory events during the cell cycle and transcription. Alterations in at least one CDK regulator or effector have been identified in almost all types of cancer. CDK8, as a member of the CDK family, serves a crucial role in gene transcription. Apart from this, phosphatidylinositol-3-kinase (PI3K)/protein kinase B (AKT) signaling pathways also play an important role in many aspects of cell growth and survival under both physiological and pathological conditions. Dysregulation of this pathway has been observed a various transformed cells and cancer tumors. In addition, aberrant activation of the Wnt/β-catenin signaling pathway causes β-catenin accumulation in the nucleus and can induce breast cancer. However, Wu et al. demonstrated that capsaicin inhibited breast cancer cell viability, induced G2/M cell cycle arrest, reduced CDK8 expression levels, decreased the phosphorylation of PI3K and Akt, and downregulated Wnt and β catenin expression levels in MDA MB 231 cells [69].

3.1.7 Bladder cancer

Bladder cancer is a common malignancy affecting the genitourinary system. It is generally subdivided into two types: nonmuscle invasive bladder cancer (NMIBC) and muscle-invasive bladder cancer (MIBC). About 30% of total patients are MIBC and have a high mortality rate due to distant metastases. Meanwhile, 70% of patients are NMIBC, which are likely to progress MIBC. Morphologically, bladder tumors can be divided into papillary, solid, and mixed types. However, the papillary type is predominant, especially in NMIBC [70]. A poor prognosis and resistance to chemotherapy are the two most important characteristics of this disease. Recently, Yang et al. reported capsaicin-induced cell death in human bladder cancer T24 cells through calcium entry-dependent ROS production and mitochondrial depolarization [71]. Likewise, Chen et al. also demonstrated capsaicin-induced cell cycle arrest by inhibiting cyclin-dependent-kinase in 5637 bladder carcinoma cells [72].

3.2 Antimicrobial activity

Microorganisms are liable for causing food spoilage and various foodborne illnesses every year. These illnesses can generate many ailments, ranging from stomach discomfort to spontaneous abortions in pregnant women, and can even lead to death in severe cases. Researchers have recently stated that some varieties of chili peppers and their active compounds exhibit significant antimicrobial properties, equivalent to some modern-day antibiotics [73–76]. Especially, Goci et al. investigated the carbopol-based formulated capsaicin enhances the antibacterial and antifungal effects against *Escherichia coli* (ATCC 10536), *Bacillus cereus* (Peru MycA 4), *Salmonella typhi* (Peru Myc 7), *Staphylococcus aureus* (ATCC 6538), *Candida tropicalis* (YEPGA 6184), *Candida albicans* (YEPGA 6379), *Candida parapsilosis* (YEPG 6551), and *Candida albicans* (YEPGA 6183), with a minimum inhibitory concentration (MIC) value reduction of at least 50% [77].

3.3 Antidiabetic activity

Diabetes mellitus is a chronic endocrine disease characterized by disorders in the metabolism of carbohydrates, lipids, and proteins due to a deficiency in insulin production by pancreatic beta cells and/or an increase in insulin resistance in peripheral tissues. Universally, this illness affects the majority of people in both developed and developing countries. Numerous synthetic drugs have been developed for the treatment, but a safe and effective paradigm is yet to be achieved. In terms of potential as a pharmacological alternative, chili has shown good antidiabetic effect because it contains α-amylase and α-glucosidase inhibitors, which are required for the degradation of polysaccharides and disaccharides. Especially, the species *Capsicum frutescens* (cayenne pepper) is often used as remedies for diabetes mellitus in African traditional medicine. Based on a previous report, Islam and Choi state that *Capsicum frutescens* increased serum insulin concentration in high-fat (HF) diet-fed streptozotocin-induced type 2 diabetes rats after four weeks of treatment. The data of this study suggest that 2% of dietary *Capsicum frutescens* is insulinotropic rather than hypoglycemic in the experimental methods [78].

3.4 Antiarthritis activity

Arthritis is an autoimmune disorder that causes pain, swelling, and stiffness in the joints. There are at least 100 types of arthritis, commonly known as connective tissue disorders, which can affect people of all ages, gender, and races. However, there is significant evidence to suggest that both the elderly and women are greatly affected. More than three decades ago, capsaicin was first shown to have protective effects in experimental arthritis [79]. Further, Inman et al. observed that capsaicin concomitantly administered with methylated bovine serum albumin (mBSA) into the rat knee markedly reduced the severity of arthritis in comparison with the contralateral inflamed knee treated with vehicle, supporting a protective role for capsaicin in reducing the severity of antigen-induced arthritis in felines [80]. According to the trend, capsaicin cream is used to reduce pain caused by many types of arthritis. Specifically, it works by decreasing a certain natural substance in the human body (substance P) that helps transmit pain signals to the brain.

3.5 Antioxidant activity

The family of free radicals generated from the oxygen is referred to as reactive oxygen species (ROS), which cause damage to other molecules by extracting electrons from them in order to attain stability. ROS are various forms of activated oxygen which include free radicals such as superoxide anion radicals (O_2^-), hydroxyl radicals (OH^-), non-free radicals (H_2O_2), and singlet oxygen [81]. The molecular basis of many diseases is known to involve oxidative stress caused by free radicals [82]. Recently progressive research has been directed at natural antioxidants. By using the DPPH free radical assay, Dubey et al. evaluated the antioxidant potential and free radical scavenging activities of some selective chili genotypes from the North East region of India in terms of inhibitory concentration (IC_{50}), efficiency concentration (EC_{50}), and anti-radical power (ARP) [83]. Likewise, Ayob et al. also determined the antioxidant activity of three varieties of Himalayan red chili (Kashmiri Local, Kupwari Local, and Shalimar Long) in North India by using DPPH radical scavenging activity, Hydroxyl radical scavenging activity, and Ferric reducing power, based on EC_{50} values [84].

3.6 Cardioprotective activity

Cardiovascular disease remains a leading cause of disability and premature death globally. The disease is mimicked by the narrowed lumen of arteries and reduced blood flow to the heart. According to a report presented at the American Heart Association's Scientific Sessions 2020, regular intake of chili peppers could significantly reduce the risk of dying from cardiovascular diseases. Moreover, very recent striking findings of a pooled longitudinal analysis by Bonaccio et al. illustrated the cardiovascular benefits of *Capsicum annuum*. This study was conducted on an Italian Cohort comprising over 22,000 men, and women originally enrolled in the Moli-sani study cohort. The results demonstrated that the use of *Capsicum annuum* is associated with a lower risk of cardiovascular mortality [85]. Most importantly, the cardiovascular system is known to be rich in capsaicin-sensitive sensory nerves, which suggests that capsaicin may regulate cardiovascular function [86]. Also, capsaicin could protect against heart disease via a transient receptor potential vanilloid subfamily member 1 (TRPV1)-mediated modulation of coronary blood flow [87]. Furthermore, the antioxidant and antiplatelet properties of capsaicin and its function in regulating energy metabolism may also contribute to its beneficial effects on the cardiovascular system [88, 89].

3.7 Neuroprotective activity

Neurodegenerative diseases such as Alzheimer's disease are often characterized by multifactorial clinical features such as loss of memory function, protein aggregation, progressive loss of neurons, cognitive impairment, neuronal cell dysfunction, and/or death. Capsaicin is a vanilloid agonist known to activate the TRPV1, recently reported to be involved in neurodegeneration [90]. A study by Veldhuis et al. demonstrated that capsaicin and the vanilloid antagonist capsazepine, peripherally administered, have been shown to exhibit neuroprotection against ouabain-induced excitotoxicity in rats [91]. Moreover, in the latest evidence, Abdel-Salam et al. reported the effect of the TRPV1 agonist capsaicin on epileptic seizures, neuronal injury, and brain oxidative stress in a model of status epilepticus induced in the rat by intraperitoneal (i.p.)

injections of pentylenetetrazole (PTZ). This study shows that 2 mg/kg of capsaicin decreased brain oxidative stress, the severity of seizures and neuronal injury, and its coadministration with phenytoin afforded neuronal protection [92].

4. Conclusion

Capsicum species is a good resource of healthy food, mainly including proteins, trace elements, vitamins, minerals, and other substances. In recent years, with the continuous enhancement of people's pursuit of nutritious, healthy food, and increasing health care awareness, chili pepper has been developed as a medicinal health supplement, functional food, and cosmetic component, especially in the dietary aspect. Moreover, its active phytochemical constituents represent a key role in treatment and disease prevention by modulating various cellular pathways. Therefore, we hope this comprehensive and updated information on chili will help promote human-based studies to facilitate its use in human health and treat diseases in the future.

Conflict of interest

The authors declare no conflict of interest.

Medical disclaimer

The authors declare that the chapter's content is for informational or educational purposes only and does not substitute professional medical advice or consultations with healthcare professionals.

Author details

Kalaiyarasi Dhamodharan, Manobharathi Vengaimaran and Mirunalini Sankaran*
Faculty of Science, Department of Biochemistry and Biotechnology, Annamalai University, Chidambaram, Tamil Nadu, India

*Address all correspondence to: mirunasankar@gmail.com

References

[1] Gomez-Garcia MD, Ochoa-Alejo N. Biochemistry and molecular biology of carotenoid biosynthesis in chili peppers (*Capsicum* spp.). International Journal of Molecular Sciences. 2013;**14**(9):19025-19053. DOI: 10.3390/ijms140919025

[2] Carrizo Garcia C, Barfuss MH, Sehr EM, Barboza GE, Samuel R, Moscone EA, et al. Phylogenetic relationships, diversification and expansion of chili peppers (*Capsicum*, Solanaceae). Annals of Botany. 2016;**118**(1):35-51. DOI: 10.1093/aob/mcw079

[3] Saxena A, Raghuwanshi R, Gupta VK, Singh HB. Chilli anthracnose: The epidemiology and management. Frontiers in Microbiology. 2016;**30**(7):1-18

[4] Paran I, Van Der Knaap E. Genetic and molecular regulation of fruit and plant domestication traits in tomato and pepper. Journal of Experimental Botany. 2007;**58**(14):3841-3852. DOI: 10.1093/jxb/erm257

[5] Food and Agriculture Organization (FAO). Food and Agriculture Organization of the United States. 2019. Available from: http://www.fao.org/faostat/en

[6] Litoriya NS, Gandhi K, Talati JG. Nutritional composition of different chilli (*Capsicum annuum* L.) varieties. Indian Journal of Agricultural Biochemistry. 2014;**27**(1):91-92

[7] Olatunji TL, Afolayan AJ. The suitability of chili pepper (*Capsicum annuum* L.) for alleviating human micronutrient dietary deficiencies: A review. Food Science & Nutrition. 2018;**6**(8):2239-2251. DOI: 10.1002/fsn3.790

[8] Khan FA, Mahmood T, Ali M, Saeed A, Maalik A. Pharmacological importance of an ethnobotanical plant: *Capsicum annuum* L. Natural Product Research. 2014;**28**(16):1267-1274. DOI: 10.1080/14786419.2014.895723

[9] Maji AK, Banerji P. Phytochemistry and gastrointestinal benefits of the medicinal spice, *Capsicum annuum* L. (Chilli): A review. Journal of Complementary and Integrative Medicine. 2016;**13**(2):97-122. DOI: 10.1515/jcim-2015-0037

[10] Kobata K, Sutoh K, Todo T, Yazawa S, Iwai K, Watanabe T. Nordihydrocapsiate, a new capsinoid from the fruits of a nonpungent pepper, *Capsicum annuum*. Journal of Natural Products. 1999;**62**(2):335-336. DOI: 10.1021/np9803373

[11] Korel F, Bagdatlioglu N, Balaban MO, Hisil Y. Ground red peppers: Capsaicinoids content, scoville scores, and discrimination by an electronic nose. Journal of Agricultural and Food Chemistry. 2002;**50**(11):3257-3261. DOI: 10.1021/jf010537b

[12] Reyes-Escogido MD, Gonzalez-Mondragon EG, Vazquez-Tzompantzi E. Chemical and pharmacological aspects of capsaicin. Molecules. 2011;**16**(2):1253-1270. DOI: 10.3390/molecules16021253

[13] Kalaiyarasi D, Mirunalini S. Capsaicin (*Capsicum annuum*): A ubiquitous compound with multivarient pharmaceutical properties. Research Journal of Chemistry and Environment. 2021;**25**(5):234-240

[14] Fattori V, Hohmann MS, Rossaneis AC, Pinho-Ribeiro FA, Verri WA. Capsaicin: Current understanding of its mechanisms and therapy of pain and other pre-clinical and clinical

uses. Molecules. 2016;**21**(7):1-33. DOI: 10.3390/molecules21070844

[15] Wang Y, Xia Y, Wang J, Luo F, Huang Y. Capsaicinoids in chili pepper (*Capsicum annuum* L.) powder as affected by heating and storage methods. Transactions of the ASABE. 2009;**52**(6):2007-2010. DOI: 10.13031/2013.29197

[16] Krithika V, Sri SR. Physicochemical and nutritional characteristics of chilli cultivars. International Journal of Research in Science. 2014;**1**(2):117-123. DOI: 10.15613/sijrs%2F2014%2Fv1i2%2F67550

[17] Vera-Guzman AM, Chavez-Servia JL, Carrillo-Rodriguez JC, Lopez MG. Phytochemical evaluation of wild and cultivated pepper (*Capsicum annuum* L. and *C. pubescens* Ruiz & Pav.) from Oaxaca, Mexico. Chilean Journal of Agricultural Research. 2011;**71**(4):578-585. DOI: 10.4067/S0718-58392011000400013

[18] Howard LR, Talcott ST, Brenes CH, Villalon B. Changes in phytochemical and antioxidant activity of selected pepper cultivars (*Capsicum* species) as influenced by maturity. Journal of Agricultural and Food Chemistry. 2000;**48**(5):1713-1720. DOI: 10.1021/jf990916t

[19] Madala N, Nutakki MK. Hot pepper-history-health and dietary benefits & production. International Journal of Current Microbiology and Applied Sciences. 2020;**9**(4):2532-2538. DOI: 10.20546/ijcmas.2020.904.303

[20] Kalaiyarasi D, Manobharathi V, Mirunalini S. Capsaicin encapsulated chitosan nanoparticles augments anticarcinogenic and antiproliferative competency against 7,12-dimethylbenz(a) anthracene induced experimental rat mammary carcinogenesis. Journal of Pharmaceutical Research International. 2021;**33**(41A):126-144. DOI: 10.9734/jpri/2021/v33i41A32311

[21] Mueller M, Hobiger S, Jungbauer A. Anti-inflammatory activity of extracts from fruits, herbs and spices. Food Chemistry. 2010;**122**(4):987-996. DOI: 10.1016/j.foodchem.2010.03.041

[22] Magied MM, Salama NA, Ali MR. Hypoglycemic and hypocholesterolemia effects of intragastric administration of dried red chili pepper (*Capsicum annum*) in alloxan-induced diabetic male albino rats fed with high-fat-diet. Journal of Food and Nutrition Research. 2014;**2**(11):850-856. DOI: 10.12691/jfnr-2-11-15

[23] Varghese S, Kubatka P, Rodrigo L, Gazdikova K, Caprnda M, Fedotova J, et al. Chili pepper as a body weight-loss food. International Journal of Food Sciences and Nutrition. 2017;**68**(4): 392-401. DOI: 10.1080/09637486.2016.1258044

[24] Shi Z, Riley M, Brown A, Page A. Chilli intake is inversely associated with hypertension among adults. Clinical Nutrition ESPEN. 2018;**23**:67-72. DOI: 10.1016/j.clnesp.2017.12.007

[25] Patcharatrakul T, Gonlachanvit S. Chili peppers, curcumins, and prebiotics in gastrointestinal health and disease. Current Gastroenterology Reports. 2016;**18**(4):1-11. DOI: 10.1007/s11894-016-0494-0

[26] Siegel RL, Miller KD, Jemal A. Cancer statistics, 2020. CA: a Cancer Journal for Clinicians. 2020;**70**(1):7-30. DOI: 10.3322/caac.21590

[27] Brenner E, Schorg BF, Ahmetlic F, Wieder T, Hilke FJ, Simon N, Schroeder C, Demidov G, Riedel T, Fehrenbacher B, Schaller M. Cancer

immune control needs senescence induction by interferon-dependent cell cycle regulator pathways in tumours. Nature Communications 2020;11(1):1-9. DOI: 10.1038/s41467-020-14987-6

[28] Diaz-Laviada I, Rodriguez-Henche N. The potential antitumor effects of capsaicin. Capsaicin as a Therapeutic Molecule. 2014;**1**:181-208. DOI: 10.1007/978-3-0348-0828-68

[29] Zhang JH, Lai FJ, Chen H, Luo J, Zhang RY, Bu HQ, et al. Involvement of the phosphoinositide 3-kinase/Akt pathway in apoptosis induced by capsaicin in the human pancreatic cancer cell line PANC-1. Oncology Letters. 2013;**5**(1):43-48. DOI: 10.3892/ol.2012.991

[30] Lee SH, Clark R. Anti-tumorigenic effects of capsaicin in colon cancer. Journal of Food Chemistry & Nanotechnology. 2016;**2**(4):162-167. DOI: 10.17756/jfcn.2016-025

[31] Scheau C, Badarau IA, Caruntu C, Mihai GL, Didilescu AC, Constantin C, et al. Capsaicin: Effects on the pathogenesis of hepatocellular carcinoma. Molecules. 2019;**24**(13):1-17. DOI: 10.3390/molecules24132350

[32] Chakraborty S, Adhikary A, Mazumdar M, Mukherjee S, Bhattacharjee P, Guha D, et al. Capsaicin-induced activation of p53-SMAR1 auto-regulatory loop down-regulates VEGF in non-small cell lung cancer to restrain angiogenesis. PLoS One. 2014;**9**(6):1-11. DOI: 10.1371/journal.pone.0099743

[33] Zhu M, Yu X, Zheng Z, Huang J, Yang X, Shi H. Capsaicin suppressed activity of prostate cancer stem cells by inhibition of Wnt/β-catenin pathway. Phytotherapy Research. 2020;**34**(4):817-824. DOI: 10.1002/ptr.6563

[34] Ferreira AK, Tavares MT, Pasqualoto KF, de Azevedo RA, Teixeira SF, Ferreira-Junior WA, et al. RPF151, a novel capsaicin-like analogue: *In vitro* studies and *in vivo* preclinical antitumor evaluation in a breast cancer model. Tumor Biology. 2015;**36**(9):7251-7267. DOI: 10.1007/s13277-015-3441-z

[35] Lin MH, Lee YH, Cheng HL, Chen HY, Jhuang FH, Chueh PJ. Capsaicin inhibits multiple bladder cancer cell phenotypes by inhibiting tumor-associated NADH oxidase (tNOX) and sirtuin1 (SIRT1). Molecules. 2016;**21**(7):1-14. DOI: 10.3390/molecules21070849

[36] Islam SU, Ahmed MB, Ahsan H, Islam M, Shehzad A, Sonn JK, et al. An update on the role of dietary phytochemicals in human skin cancer: New insights into molecular mechanisms. Antioxidants. 2020;**9**(10):1-30. DOI: 10.3390/antiox9100916

[37] Clark R, Lee SH. Anticancer properties of capsaicin against human cancer. Anticancer Research. 2016;**36**(3):837-843

[38] Lin S, Zhang J, Chen H, Chen K, Lai F, Luo J, et al. Involvement of endoplasmic reticulum stress in capsaicin-induced apoptosis of human pancreatic cancer cells. Evidence-based Complementary and Alternative Medicine. 2013;**1**:1-12. DOI: 10.1155/2013/629750

[39] Pramanik KC, Srivastava SK. Apoptosis signal-regulating kinase 1–thioredoxin complex dissociation by capsaicin causes pancreatic tumor growth suppression by inducing apoptosis. Antioxidants & Redox Signaling. 2012;**17**(10):1417-1432. DOI: 10.1089/ars.2011.4369

[40] Kim CS, Park WH, Park JY, Kang JH, Kim MO, Kawada T, et al. Capsaicin, a spicy component of hot pepper, induces apoptosis by activation of the peroxisome

proliferator-activated receptor γ in HT-29 human colon cancer cells. Journal of Medicinal Food. 2004;**7**(3):267-273. DOI: 10.1089/jmf.2004.7.267

[41] Jin J, Lin G, Huang H, Xu D, Yu H, Ma X, et al. Capsaicin mediates cell cycle arrest and apoptosis in human colon cancer cells via stabilizing and activating p53. International Journal of Biological Sciences. 2014;**10**(3):285-295. DOI: 10.7150/ijbs.7730

[42] Lu HF, Chen YL, Yang JS, Yang YY, Liu JY, Hsu SC, et al. Antitumor activity of capsaicin on human colon cancer cells *in vitro* and colo 205 tumor xenografts *in vivo*. Journal of Agricultural and Food Chemistry. 2010;**58**(24):12999-13005. DOI: 10.1021/jf103335w

[43] Wang KX, Fang JS, Qin XM, Du GH, Gao L. Uncovering the anti-metastasis effects and mechanisms of capsaicin against hepatocellular carcinoma cells by metabolomics. Journal of Functional Foods. 2019;**60**:1-9. DOI: 10.1016/j.jff.2019.103431

[44] Brown KC, Witte TR, Hardman WE, Luo H, Chen YC, Carpenter AB, et al. Capsaicin displays anti-proliferative activity against human small cell lung cancer in cell culture and nude mice models via the E2F pathway. PLoS One. 2010;**5**(4):1-15. DOI: 10.1371/journal.pone.0010243

[45] Ramos-Torres A, Bort A, Morell C, Rodriguez-Henche N, Diaz-Laviada I. The pepper's natural ingredient capsaicin induces autophagy blockage in prostate cancer cells. Oncotarget. 2016;**7**(2):1569-1583. DOI: 10.18632/oncotarget.6415

[46] Venier NA, Yamamoto T, Sugar LM, Adomat H, Fleshner NE, Klotz LH, et al. Capsaicin reduces the metastatic burden in the transgenic adenocarcinoma of the mouse prostate model. The Prostate. 2015;**75**(12):1300-1311. DOI: 10.1002/pros.23013

[47] Chang HC, Chen ST, Chien SY, Kuo SJ, Tsai HT, Chen DR. Capsaicin may induce breast cancer cell death through apoptosis-inducing factor involving mitochondrial dysfunction. Human & Experimental Toxicology. 2011;**30**(10):1657-1665. DOI: 10.1177/0960327110396530

[48] Chen M, Xiao C, Jiang W, Yang W, Qin Q, Tan Q, et al. Capsaicin inhibits proliferation and induces apoptosis in breast cancer by down-regulating FBI-1-mediated NF-κB pathway. Drug Design, Development and Therapy. 2021;**15**:125-140. DOI: 10.2147/DDDT.S269901

[49] Qian K, Wang G, Cao R, Liu T, Qian G, Guan X, et al. Capsaicin suppresses cell proliferation, induces cell cycle arrest and ROS production in bladder cancer cells through FOXO3a-mediated pathways. Molecules. 2016;**21**(10):1-15. DOI: 10.3390/molecules21101406

[50] Hidalgo M, Cascinu S, Kleeff J, Labianca R, Lohr JM, Neoptolemos J, et al. Addressing the challenges of pancreatic cancer: Future directions for improving outcomes. Pancreatology. 2015;**15**(1):8-18. DOI: 10.1016/j.pan.2014.10.001

[51] Pramanik KC, Boreddy SR, Srivastava SK. Role of mitochondrial electron transport chain complexes in capsaicin mediated oxidative stress leading to apoptosis in pancreatic cancer cells. PLoS One. 2011;**6**(5):1-16. DOI: 10.1371/journal.pone.0020151

[52] Pramanik KC, Fofaria NM, Gupta P, Ranjan A, Kim SH, Srivastava SK. Inhibition of β-catenin signaling suppresses pancreatic tumor growth by disrupting nuclear β-catenin/

TCF-1 complex: Critical role of STAT-3. Oncotarget. 2015;**6**(13):11561-11574. DOI: 10.18632/oncotarget.3427

[53] Sung H, Ferlay J, Siegel RL, Laversanne M, Soerjomataram I, Jemal A, et al. Global cancer statistics 2020: GLOBOCAN estimates of incidence and mortality worldwide for 36 cancers in 185 countries. CA: a Cancer Journal for Clinicians. 2021;**71**(3):209-249. DOI: 10.3322/caac.21660

[54] Kumar R. Commentary: Targeting colorectal cancer through molecular biology. Seminars in Oncology. 2005;**32**:37-39. DOI: 10.1053/j.seminoncol.2005.06.012

[55] Giannakis M, Mu XJ, Shukla SA, Qian ZR, Cohen O, Nishihara R, et al. Genomic correlates of immune-cell infiltrates in colorectal carcinoma. Cell Reports. 2016;**15**(4):857-865. DOI: 10.1016/j.celrep.2016.03.075

[56] Bessler H, Djaldetti M. Capsaicin modulates the immune cross talk between human mononuclears and cells from two colon carcinoma lines. Nutrition and Cancer. 2017;**69**(1):14-20. DOI: 10.1080/01635581.2017.1247893

[57] Dhanasekaran R, Felsher DW. A tale of two complications of obesity: Nonalcoholic steatohepatitis (NASH) and hepatocellular carcinoma (HCC). Hepatology (Baltimore, MD.). 2019;**70**(3):1056-1058. DOI: 10.1002/hep.30649

[58] Huang SP, Chen JC, Wu CC, Chen CT, Tang NY, Ho YT, et al. Capsaicin-induced apoptosis in human hepatoma HepG2 cells. Anticancer Research. 2009;**29**(1):165-174

[59] Bort A, Spinola E, Rodriguez-Henche N, Diaz-Laviada I. Capsaicin exerts synergistic antitumor effect with sorafenib in hepatocellular carcinoma cells through AMPK activation. Oncotarget. 2017;**8**(50):87684-87698. DOI: 10.18632/oncotarget.21196

[60] Gaspar MM, Radomska A, Gobbo OL, Bakowsky U, Radomski MW, Ehrhardt C. Targeted delivery of transferrin-conjugated liposomes to an orthotopic model of lung cancer in nude rats. Journal of Aerosol Medicine and Pulmonary Drug Delivery. 2012;**25**(6):310-318. DOI: 10.1089/jamp.2011.0928

[61] Han TH, Park MK, Nakamura H, Ban HS. Capsaicin inhibits HIF-1α accumulation through suppression of mitochondrial respiration in lung cancer cells. Biomedicine & Pharmacotherapy. 2022;**146**:1-7. DOI: 10.1016/j.biopha.2021.112500

[62] Lau JK, Brown KC, Dom AM, Witte TR, Thornhill BA, Crabtree CM, et al. Capsaicin induces apoptosis in human small cell lung cancer via the TRPV6 receptor and the calpain pathway. Apoptosis. 2014;**19**(8):1190-1201. DOI: 10.1007/s10495-014-1007-y

[63] Taplin ME, Ho SM. The endocrinology of prostate cancer. The Journal of Clinical Endocrinology & Metabolism. 2001;**86**(8):3467-3477. DOI: 10.1210/jcem.86.8.7782

[64] Jansson MD, Lund AH. MicroRNA and cancer. Molecular Oncology. 2012;**6**(6):590-610. DOI: 10.1016/j.molonc.2012.09.006

[65] Zheng L, Chen J, Ma Z, Liu W, Yang F, Yang Z, et al. Capsaicin causes inactivation and degradation of the androgen receptor by inducing the restoration of miR-449a in prostate cancer. Oncology Reports. 2015;**34**(2):1027-1034. DOI: 10.3892/or.2015.4055

[66] Kim SM, Oh EY, Lee JH, Nam D, Lee SG, Lee J, et al. Brassinin combined with capsaicin enhances apoptotic and anti-metastatic effects in PC-3 human prostate cancer cells. Phytotherapy Research. 2015;**29**(11):1828-1836. DOI: 10.1002/ptr.5478

[67] Sanchez BG, Bort A, Mateos-Gomez PA, Rodriguez-Henche N, Diaz-Laviada I. Combination of the natural product capsaicin and docetaxel synergistically kills human prostate cancer cells through the metabolic regulator AMP-activated kinase. Cancer Cell International. 2019;**19**(1):1-4. DOI: 10.1186/s12935-019-0769-2

[68] Thoennissen NH, O'kelly J, Lu D, Iwanski GB, La Abbassi S, Leiter A, et al. Capsaicin causes cell-cycle arrest and apoptosis in ER-positive and-negative breast cancer cells by modulating the EGFR/HER-2 pathway. Oncogene. 2010;**29**(2):285-296. DOI: 10.1038/onc.2009.335

[69] Wu D, Jia H, Zhang Z, Li S. Capsaicin suppresses breast cancer cell viability by regulating the CDK8/PI3K/Akt/Wnt/β-catenin signaling pathway. Molecular Medicine Reports. 2020;**22**(6):4868-4876. DOI: 10.3892/mmr.2020.11585

[70] Kamat AM, Hahn NM, Efstathiou JA, Lerner SP, Malmström P-U, Choi W, et al. Bladder cancer. Lancet. 2016;**388**(10061):2796-2810. DOI: 10.1016/S0140-6736(16)30512-8

[71] Yang ZH, Wang XH, Wang HP, Hu LQ, Zheng XM, Li SW. Capsaicin mediates cell death in bladder cancer T24 cells through reactive oxygen species production and mitochondrial depolarization. Urology. 2010;**75**(3):735-741. DOI: 10.1016/j.urology.2009.03.042

[72] Chen D, Yang Z, Wang Y, Zhu G, Wang X. Capsaicin induces cycle arrest by inhibiting cyclin-dependent-kinase in bladder carcinoma cells. International Journal of Urology. 2012;**19**(7):662-668. DOI: 10.1111/j.1442-2042.2012.02981.x

[73] Raybaudi-Massilia R, Suarez AI, Arvelo F, Zambrano A, Sojo F, Calderon-Gabaldon MI, et al. Cytotoxic, antioxidant and antimicrobial properties of red sweet pepper (*Capsicum annuum* L. Var. Llaneron) extracts: *In vitro* study. International Journal of Food Studies. 2017;**6**(2):222-231. DOI: 10.7455/ijfs/6.2.2017.a8

[74] Gurnani N, Gupta M, Mehta D, Mehta BK. Chemical composition, total phenolic and flavonoid contents, and *in vitro* antimicrobial and antioxidant activities of crude extracts from red chilli seeds (*Capsicum frutescens* L.). Journal of Taibah University for Science. 2016;**10**(4):462-470. DOI: 10.1016/j.jtusci.2015.06.011

[75] Sen N, Paul D, Sinha SN. *In vitro* antibacterial potential and phytochemical analysis of three species of chilli plant. Journal of Chemical and Pharmaceutical Research. 2016;**8**(2):443-447

[76] Marini E, Magi G, Mingoia M, Pugnaloni A, Facinelli B. Antimicrobial and anti-virulence activity of capsaicin against erythromycin-resistant, cell-invasive group A Streptococci. Frontiers in Microbiology. 2015;**6**:1-7. DOI: 10.3389/fmicb.2015.01281

[77] Goci E, Haloci E, Di Stefano A, Chiavaroli A, Angelini P, Miha A, et al. Evaluation of *in vitro* capsaicin release and antimicrobial properties of topical pharmaceutical formulation. Biomolecules. 2021;**11**(3):1-10. DOI: 10.3390/biom11030432

[78] Islam MS, Choi H. Dietary red chilli (*Capsicum frutescens* L.) is insulinotropic rather than hypoglycemic in type 2

diabetes model of rats. Phytotherapy Research. 2008;**22**(8):1025-1029. DOI: 10.1002/ptr.2417

[79] Colpaert FC, Donnerer J, Lembeck F. Effects of capsaicin on inflammation and on the substance P content of nervous tissues in rats with adjuvant arthritis. Life Sciences. 1983;**32**(16):1827-1834. DOI: 10.1016/0024-3205(83)90060-7

[80] Inman RD, Chiu B, Rabinovich S, Marshall W. Neuromodulation of synovitis: Capsaicin effect on severity of experimental arthritis. Journal of Neuroimmunology. 1989;**24**(1-2):17-22. DOI: 10.1016/0165-5728(89)90093-3

[81] Halliwell B. How to characterize an antioxidant: An update. Biochemical Society Symposium. 1995;**61**:73-101. DOI: 10.1042/bss0610073

[82] Cross CE, Halliwell B, Borish ET, Pryor WA, Ames BN, Saul RL, et al. Oxygen radicals and human disease. Annals of Internal Medicine. 1987;**107**(4):526-545. DOI: 10.7326/0003-4819-107-4-526

[83] Dubey RK, Singh V, Upadhyay G, Pandey AK, Prakash D. Assessment of phytochemical composition and antioxidant potential in some indigenous chilli genotypes from North East India. Food Chemistry. 2015;**188**:119-125. DOI: 10.1016/j.foodchem.2015.04.088

[84] Ayob O, Hussain PR, Suradkar P, Naqash F, Rather SA, Joshi S, et al. Evaluation of chemical composition and antioxidant activity of Himalayan Red chilli varieties. LWT. 2021;**146**:1-9. DOI: 10.1016/j.lwt.2021.111413

[85] Bonaccio M, Di Castelnuovo A, Costanzo S, Ruggiero E, De Curtis A, Persichillo M, et al. Chili pepper consumption and mortality in Italian adults. Journal of the American College of Cardiology. 2019;**74**(25):3139-3149. DOI: 10.1016/j.jacc.2019.09.068

[86] Vaishnava P, Wang DH. Capsaicin sensitive-sensory nerves and blood pressure regulation. Current Medicinal Chemistry. Cardiovascular and Hematological Agents. 2003;**1**(2):177-188. DOI: 10.2174/1568016033477540

[87] Guarini G, Ohanyan VA, Kmetz JG, DelloStritto DJ, Thoppil RJ, Thodeti CK, et al. Disruption of TRPV1-mediated coupling of coronary blood flow to cardiac metabolism in diabetic mice: Role of nitric oxide and BK channels. American Journal of Physiology. Heart and Circulatory Physiology. 2012;**303**(2):216-223. DOI: 10.1152/ajpheart.00011.2012

[88] Luo XJ, Peng J, Li YJ. Recent advances in the study on capsaicinoids and capsinoids. European Journal of Pharmacology. 2011;**650**(1):1-7. DOI: 10.1016/j.ejphar.2010.09.074

[89] Juturu V. Capsaicinoids modulating cardiometabolic syndrome risk factors: Current perspectives. Journal of Nutrition and Metabolism. 2016;**2016**:1-11. DOI: 10.1155/2016/4986937

[90] Mezey E, Toth ZE, Cortright DN, Arzubi MK, Krause JE, Elde R, et al. Distribution of mRNA for vanilloid receptor subtype 1 (VR1), and VR1-like immunoreactivity, in the central nervous system of the rat and human. Proceedings of the National Academy of Sciences. 2000;**97**(7):3655-3660. DOI: 10.1073/pnas.97.7.3655

[91] Veldhuis WB, Van der Stelt M, Wadman MW, Van Zadelhoff G, Maccarrone M, Fezza F, et al. Neuroprotection by the endogenous cannabinoid anandamide and arvanil against *in vivo* excitotoxicity in the rat: Role of vanilloid receptors and

lipoxygenases. Journal of Neuroscience. 2003;**23**(10):4127-4133. DOI: 10.1523/JNEUROSCI.23-10-04127.2003

[92] Abdel-Salam OM, Sleem AA, Sayed MA, Youness ER, Shaffie N. Capsaicin exerts anti-convulsant and neuroprotective effects in pentylenetetrazole-induced seizures. Neurochemical Research. 2020;**45**(5):1045-1061. DOI: 10.1007/s11064-020-02979-3

Chapter 5

Phytophthora capsici on *Capsicum* Plants: A Destructive Pathogen in Chili and Pepper Crops

Anthony A. Moreira-Morrillo,
Álvaro Monteros-Altamirano, Ailton Reis
and Felipe R. Garcés-Fiallos

Abstract

Capsicum from tropical and subtropical America, is an important genus for the nutritional, economic and cultural values of its species. At the same time, the *Capsicum* species are affected by diseases caused by viruses, bacteria, fungi and pseudofungi, in particular the oomycete *Phytophthora capsici*. This phytopathogen causes great damage and losses in different *Capsicum* species, because it infects all plant organs causing root, crown and fruit rot; and, leaf blight. The polycyclic dispersion through zoospores and sporangia, the limited availability of resistant genotypes, and the reduced diversity of effective oomyceticides (fungicides), make *P. capsici* one of the most complex phytopathogens to be managed worldwide specially in field conditions. However, successful management of *P. capsici* depends on the knowledge of the pathogen, its interaction with the susceptible host and the methods of control used. Thus, this chapter addressed the etiology, symptomatology, occurrence and management of the disease. Additionally, the cycle of the disease is discussed in a holistic and simple way.

Keywords: *Capsicum* spp., root and crown rot, leaf blight, fruit rot, *Phytophthora capsici* cycle, disease management

1. Introduction

Capsicum is a genus native to tropical and subtropical America [1], where species such as *C. annuum* L., *C. baccatum* L., *C. chinense* Jacq., *C. frutescens* L., and *C. pubescens* R. & P. [2, 3] excel for its great nutritional, economic [4], and cultural value in the gastronomy of several countries [5]. These vegetables, also commonly known as sweet peppers or hot peppers, can reach a global production of 38 million tons [6]. However, like other crops, peppers are not exempt from disease attacks, caused mainly by viruses, bacteria (leaf spots and vascular wilts), fungi (cercosporiosis, powdery mildew and anthracnose), and mainly pseudofungi (rot of roots, stems and fruits; and, leaf blight) [7].

Phytophthora capsici is an oomycete present in several parts of the world, being reported as the causal agent of countless diseases in different crops of agricultural importance such as Cucurbitaceae, Rosaceae, Fabaceae, Liliaceae and Solanaceae [8]. This phytopathogen in *Capsicum* can cause damage of up to 100% due to its rapid spread in field conditions and represents around $100 million in losses [9, 10] for which it is considered the fifth most destructive oomycete in the world [11]. Due to the reach of aerial tissues, its polycyclic characteristic makes *P. capsici* one of the most complex phytopathogens to be managed [12].

Resistant plants can generally activate different biochemical, structural and molecular defense mechanisms against the infection of *P. capsici* [13, 14]; conversely, susceptible plants can be infected and colonized by the pathogen. Since most of the common commercial genotypes of *Capsicum* are susceptible, i.e. Chinese Giant (CG), California Wonder (CW), Osh Kosh (OK) and Yolo Wonder (YW), farmers use a considerable amount of oomyceticides to control its attack [3]. However, there are other measures that can be used in an integrated manner to drastically reduce infection of *P. capsici* i.e. crop rotation, irrigation management, use of biocontrole agents, among others. For example, the rate of progress of collar rot in *Capsicum* plants can be reduced considerably by applying *Trichoderma harzianum* [15, 16]. Therefore, due to the importance of *P. capsici* in *Capsicum* spp., this chapter address the etiology, symptomatology, worldwide occurrence and biological cycle of the pathogen in the different plant tissues; and, finally describes the disease management measures to be used either in isolation or integrated approach.

2. Etiology

Phytophthora capsici belongs to the Phylum: Oomycota, Class: Oomycetes, Order: Peronosporales, and Family: Peronosporaceae (**Figure 1**). Under laboratory conditions, the phytopathogen can grow in culture media based on V8 juice [17, 18], carrot, tomato, and water agar. Although the growth pattern of the oomycete can be different between isolates, the shape of the colonies is generally rounded and whitish [19], standing out those with a slightly stellate pattern (**Figure 1A**). The phytopathogen characteristics are: elongated, coenocytic hyphae (**Figure 1B**); stretched cells with several nuclei [20], measuring between 3 and 8 μm wide [21]; oospores with a diameter between 25.6 and 52.4 μm (**Figure 1B**) [22]. Its sporangia (**Figure 1C**) are abundant and formed individually or branched, flaccid or closely spaced [21], mainly spherical (globose, ellipsoid and ovoid), pear-shaped (obpiriform and obturbinate), lemon (limoniform), and/or in some cases irregular (distorted) [19], measuring between 32.1 and 51.3 μm long and 23.3–37.1 μm wide [20]. The main pedicel has a length of 38.3 to 84.4 μm [22]. The zoospores released from the sporangia (**Figure 1D**) have a diameter between 5.84 and 11.3 μm [20]. Although resistance structures such as chlamydospores can also be formed by some *Phytophthora* species [21], it is not common in *P. capsici* [18].

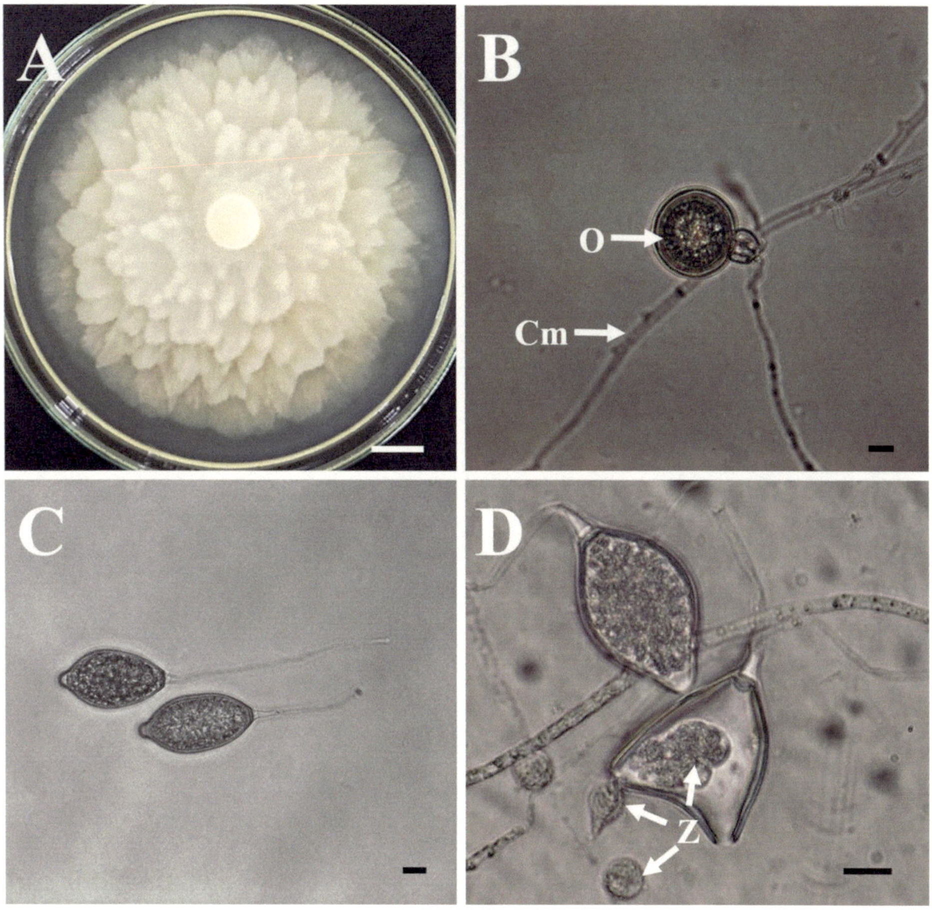

Figure 1.
Morphological structures of Phytophthora capsici*: A) Colony morphology after 14 d growth at 24 ± 3°C on potato dextrose agar; B) Coenocytic mycelium, and oospore; C) Papillate limoniform sporangia; D) Zospores inside and outside of sporangia. Cm: Coenocytic mycelium; O: Oospore; Z: Zoospores. Bars: 1 cm (A) and 10 μm (B-D). Source: Unpublished photographs from the authors.*

3. Symptomatology

Several symptoms (**Figure 2**) may be caused by *P. capsici* in different phenological stages of the plant [23], which can vary depending on host resistance, affected tissue and/or climatic conditions [24]. In susceptible plants, the initial symptoms can appear between four and seven days after inoculation (DAI), and can die at approximately 10 DAI [25, 26]. Meanwhile, in resistant plants after 10 DAI, slight lesions appear, mainly due to the ability to counteract the advance of colonization [3, 27].

The root system is the first affected tissue (**Figure 2A**), showing a brown rot [3]. Later, brown lesions with a rough shape can be observed on the crown tissues (**Figure 2A**) [28]. Other symptoms observed in adult plants are stunting (**Figure 2B**) and generalized wilting (**Figure 2C**) [9, 26, 29]. In the foliar area, leaf blight may be

Figure 2.
Symptoms caused by Phytophthora capsici *on different organs of* Capsicum *plants: A) root and crown rot; B) dwarf; C) wilting; D) leaf blight; E) fruit rot. Rc: Root crown. Rs: Root system. Source: Unpublished photographs from the authors.*

observed (**Figure 2D**) starting with small dark water-soaked lesions, which later become necrotic with a light brown center and dark edges [30, 31]. Fruits are also affected by the oomycete, initially, minute lesions with clear whitish centers are observed in the tissue [32, 33], advancing rapidly until this organ completely rots (**Figure 2E**) [18]. This last symptom is often presented in young fruits, compared to mature fruits that present more resistance to *P. capsici* [34].

4. Occurrence of *Phytophthora capsici* worldwide

Phytophthora capsici has been reported affecting various crops in several countries, however, we focus on the pathogen's reports on *Capsicum* spp. (**Figure 3**). The first report of the oomycete dates back to 1922 in New Mexico, attacking branches and fruits of chili plants [35]. Subsequently, the pathogen was reported in the Netherlands [28], Bulgaria [36], Brazil [37], South Africa [38], Korea [39] and Tunisia [40], all affecting pepper.

From 2002 onwards, the oomycete was reported affecting *Capsicum* spp. crops in different countries and regions such as Egypt [41], Spain [42], and Italy [43], China [44], Canada [45], Taiwan [46], Peru [47], Algeria [48], Mexico [49], Laos PDR [29], Pakistan, [50], Bhutan [22], Indonesia [51], Trinidad and Tobago [20] and Ecuador [52]. Despite the oomycete's reported geographical distribution affecting mainly chili and sweet peppers, some other countries could be affected, but official reports are missing [9].

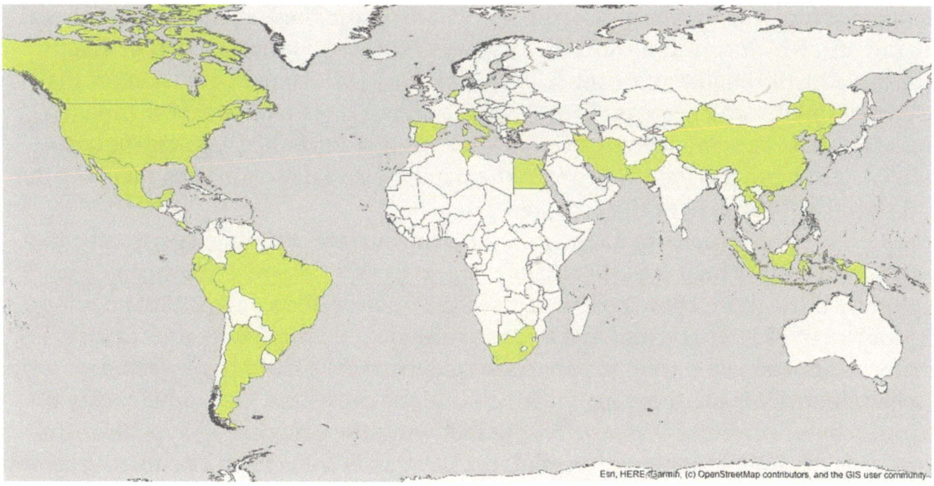

Figure 3.
Global geographic distribution of Phytophthora capsici *affecting tissues of* Capsicum *plants. This figure was developed with the help of the BioRender platform. Source: Unpublished photographs from the authors.*

5. Disease cycle of *Phytophthora capsici*

Every disease begins with the survival of the phytopathogen (source of primary inoculum); in the case of *P. capsici* (**Figure 4**) the mycelium can survive in the soil and cultural debris, or in the weeds that serve as facultative hosts. Besides, the oospores

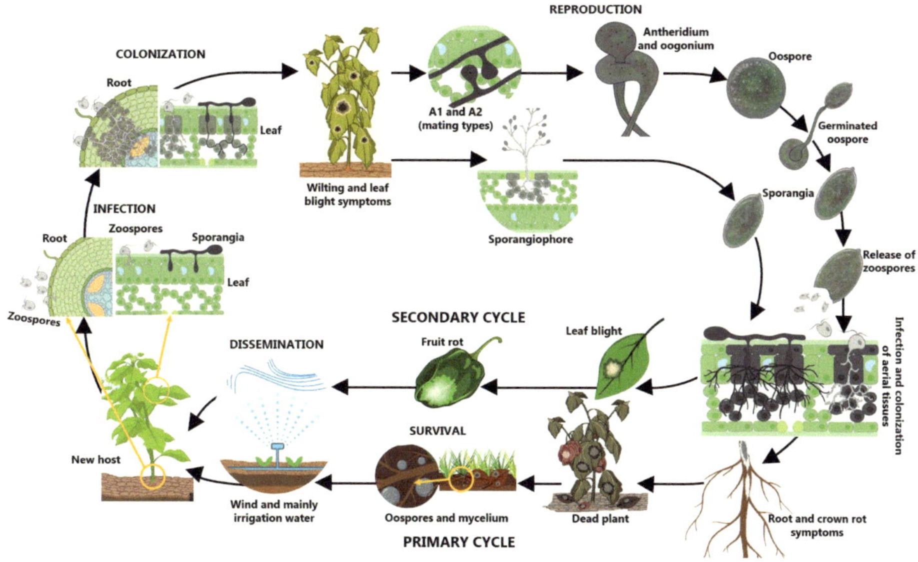

Figure 4.
Biological cycle of Phytophthora capsici *causing root, crown and fruit rot, and foliar blight in* Capsicum *plants. This figure was developed with the help of the BioRender platform. Source: Unpublished photographs from the authors.*

can persist for long periods in the soil [16, 53, 54]. From these structures, sporangia are formed, which spread towards the tissues of the host located in near or distant crops, mainly through water (rainfall or irrigation) [18]. Under conditions of high relative humidity and between 27 and 32°C, the sporangia release motile biflagellate zoospores that swim through the irrigation water until they reach plant tissues [12, 55]. Although it is mentioned that the wind is also a disseminating factor of the pathogen [56], this is not fully accepted yet.

Sporangia can germinate directly on the plant surface, forming a germ tube that penetrates the anticlinal walls of rhizodermal cells [12]. Likewise, they can do it indirectly through zoospores released from the interior of sporangia that encyst on the surface of root, crown and leaf tissues, subsequently producing a germ tube, and finally an appressorium used to penetrate epidermal cells [55–57]. *Phytophthora capsici* is a hemibiotrophic phytopathogen, which initially presents a biotrophic lifestyle, followed by a necrotrophic phase [58]. In fact, once the penetrating hyphae of the oomycete enter plant tissues intercellularly, these structures form haustoria to remove nutrients from the cells [57, 58]. Subsequently, vegetative hyphae and haustoria are formed in the form of lateral branches that colonize intracellularly and epicuticularly [27]. Finally, *P. capsici* intensely colonizes epidermal, vascular (phloem, xylem) and parenchymal cells [16]. Both the infection and colonization processes (latent period) can last between four and seven days [3].

The last phase of the primary cycle of the disease is completed with the reproduction of the pathogen, which occurs on the external surface of the host. *Phytophthora capsici* is a heterothallic species that has one of two mating types called A1 and A2, producing a male gametangium (antheridium) and a female gametangium (oogonium), resulting in sexual spores called oospores, with thick walls adapted to survive winter and unfavorable weather conditions [30, 55, 59]. These reproductive structures go through a rest period and serve for the survival of the phytopathogen [53]. There is also asexual reproduction that is characterized by the formation of sporangia from branched sporangiophores [30, 55]. The production of sporangia occurs between 25 and 30°C, under conditions of high relative humidity, and ≈90 hours after infection, which will produce zoospores by cytoplasmic cleavage [24, 55].

Although root and crown rot are monocyclic diseases, others such as leaf blight and fruit rot would be polycyclic. Thus, the sporangia of *P. capsici* would function as a secondary inoculum [53, 57]. These structures or propagules would reach the aerial tissues through water splashes, initiating a new infectious cycle [53, 60, 61], repeating the previously mentioned phases of infection, colonization and reproduction. The ability of the phytopathogen to reach practically all plant tissues makes its management complex.

6. Disease management

The management of diseases caused by *P. capsici* may be very difficult and economically expensive, especially due to the excessive use of oomyceticides (formerly called fungicides). However, there are different alternatives which used in an integrated manner during the pre-sowing, production and post-harvest stages could reduce damage and losses in *Capsicum* crops. These alternatives are: the use of resistant cultivars, well-drained soils, crop rotation, soil treatments, tillage methods, irrigation control, improvement of irrigation water quality, and use of plastic mulches [12]. In any case, the infection of different plant organs of *Capsicum* (**Figure 2**) by

P. caspici makes the integrated implementation of measures by farmers, complex but worthwhile [16, 18].

6.1 Genetic control

Obtaining resistant germplasm of *Capsicum* to *P. capsici* is a complex task, requiring different breeding techniques and germplasm screening including landraces as sources of resistance [9]. So far there are some resistant commercial cultivars such as Nathalie, Paladin, Ungara, Violeta, Ayesha, Violeta 1, Sempurna, and Ayesha Ungu, which are being used worldwide [3, 51]. Likewise, there are different resistant landraces such as CM-334, (*C. annuum*) ECU-12831 (*C. baccatum*), ECU-9129, (*C. chinense*) Code 5 (*C. frutescens*), ECU-1296 (*C. frutescens*), found in Ecuador and Mexico [3, 62] that could be used in breeding programs.

In one hand, *P. capsici* has developed different mechanisms to attack plants and to obtain its necessary nutrients, while on the other hand, plants have developed a complex defense system that prevents the entry or limits the advance of the oomycete in plant tissue, including physical, biochemical and molecular mechanisms [13, 14, 63]. One of the first barriers in pepper plants (*C. annuum*) limiting *P. capsici* infection is a thick cell wall and a high content of phenolics and flavonoids [16], such as the soluble phenols chlorogenic acid, luteolin glycoside, apiosil glucoside of luteolin, and aglycone apigenin [64, 65]. Other mechanisms include the synthesis of antimicrobial phytoalexins, the induction of hydrolytic enzymes such as chitinase and glucanase, and the production of proteins rich in hydroxyproline, reactive oxygen species (ROS), and capsidiol [66, 67]. Regarding the latter mechanism, this *Capsicum* phytoalexin could inhibit oomycete development [66, 67]. These and other mechanisms make *Capsicum* plants prevent or considerably delay the infection, colonization and reproduction of *P. capsici* in the different subterranean or aerial tissues.

6.2 Biological control

The current need to consume healthy foods, free of synthetic-pesticide residues [68], has led to the promotion of alternatives such as the use of effective biological control agents i.e. *Bacillus* spp., *Trichoderma* spp., among others, which if used under suitable climatic conditions will contribute significantly and economically to the prevention and management of diseases caused by *P. capsici*, in addition to promoting the growth of *Capsicum* [69, 70].

The use of microorganisms such as *Bacillus* spp. and *Trichoderma* spp. are highly valuable alternatives for the management of diseases caused by *P. capsici*. Under laboratory and green house conditions, *Bacillus amyloliquefaciens* (strain PsL) can reduce the mycelial growth of *P. capsici* by up to 46%, in addition to the growth promotion of *P. capsici* pepper plants [70]. *Bacillus subtilis* (isolates R13 and R33) can reduce the incidence of foliar blight between 71 and 87% [23]. *In vitro* and *in vivo* experiments of native *Trichoderma* strains against *P. capsici* isolates in *C. pubescens* plants, showed that *T. harzianum* inhibits the radial growth of the phytopathogen by 43%, and reduces plant mortality by 10% at 20 DAI [71].

Endophytic microorganisms can also be used in biological control. Some of them such as *Nigrospora sphaerica*, *Enterobacter* sp. and *Dothideales* sp. have been used as biocontrollers of pathogens affecting *C. annuum*, such as *P. capsici* [72]. *Nigrospora sphaerica* (isolate A22F1) was used to control *P capsici* in susceptible seedlings of *C. annuum* (cvs. California Wonder, Numex spring and Pepper cayene), observing a considerable reduction in root rot compared to control. Recently, a metagenomic study [73] found different fungal species that are used in the biological control of phytopathogens associated with the mycobiome of resistant and susceptible hypocotyls, infected or not with *P. capsici*.

6.3 Cultural control

Cultural control is based on the use of measures that favor the development of the crop, and at the same time, affecting the phytopathogen, in order to reduce the intensity of the disease [74]. Strategies include limiting soil saturation, water accumulation in plots, and movement of infected plant debris or infested soil within a field [18]. Crop rotation is another very important aspect to consider because it affects the survival of the phytopathogen and the host range [26] e.g. crop rotation for 3 years can considerably reduce the propagules (mainly oospores) of *P. capsici*, which have the ability to remain in the soil for long periods of time [16, 22].

6.4 Chemical control

The use of oomyceticides (previously called fungicides) is common in the management of diseases caused by *P. capsici* in *Capsicum* plants (Table 1), especially those that contain molecules with a direct mode of action on the phytopathogen [75]. The efficacy of these oomyceticides, of synthetic origin, has been demonstrated under laboratory and field conditions. For example, Mancozeb 64% + Metalaxyl 4% (7.5 g L^{-1}) or Copper sulfate pentahydrate (2.5 mL L^{-1}) applied to the soil, and Potassium phosphonate (5 mL L^{-1}) applied to the leaf area, can totally reduce the incidence of root and crown rot [47]. Also, the use of Fosetil Aluminum (2.5 kg ha^{-1}) applied to the soil (drench) can reduce up to 100% of wilting in pepper plants [76]. Other molecules such as ametoctradine + dimethomorph, cyazofamid, dimethomorph, famoxadon + cymoxanil, fluazinam, fluopicolide, mandipropamide, mefenoxam, phosphonates, and zoxamide + mancozeb, can also be used to control damping-off, leaf blight and fruit rot [26].

Despite the success achieved over the years with the use of chemical control, the inappropriate use of molecules has made some *P. capsici* isolates insensitive to commercial oomyceticides such as metalaxyl and mefenoxam [77, 78]. A solution to reduce these effects on the oomycete is the use of active ingredients such as mandipropamide and dimethomorph, considered molecules with low to medium risk of resistance [79, 80]. In order to reduce the selection pressure by the phytopathogen, the farmer must have a wide range of molecules applied in periodic and scheduled rotation during each crop cycle [81], and even use mixtures that have systemic and protective modes of action.

6.5 Integrated disease management

Integrated disease management (IDM) aims to minimize the biological activity of the pathogen and increase crop productivity, involving the use of various techniques in favor of the environment by avoiding the excessive use of chemical molecules and reducing control costs of *P. capsici* [80]. To effectively manage diseases caused by *P. capsici* in *Capsicum,* different management strategies should be integrated in either agroecological, conventional or other production systems [81]. Usually a single strategy is ineffective for the management of *P. capsici* [80]; for this, IDM is based on immunization, exclusion, eradication and crop protection mainly including soil and plant management through soil amendments, solarization, crop cover, water treatment, seed treatment (with a biofungicide to improve germination and reduce the incidence of damping-off), and others [82–84].

7. Conclusions

Without a doubt *Phytophthora capsici* is a dangerous pathogen for horticultural fields; negatively affecting both young and adult chili and sweet pepper plants. The versatility of this plant pathogen to spread in the field, and to infect and to colonize practically all the plant organs, make the management of the diseases a huge challenge. However, different control methods such as genetic, biological, cultural and chemical or integrated disease management may be integrated to mitigate damages and losses in *Capsicum* spp. production fields. Despite the advances already made around *P. capsici*, there is still much to study and learn about this phytopathogen.

Conflict of interest

The authors declare that they have no known competing financial interests or personal relationships that could have appeared to influence the work reported in this chapter.

Author details

Anthony A. Moreira-Morrillo[1], Álvaro Monteros-Altamirano[2], Ailton Reis[3] and Felipe R. Garcés-Fiallos[1]*

1 Faculty of Agronomic Engineering, Technical University of Manabí, Laboratory of Phytopathology, Experimental Campus La Teodomira, Santa Ana, Manabí, Ecuador

2 Departamento Nacional de Recursos Fitogenéticos (DENAREF), Instituto Nacional Autónomo de Investigaciones Agropecuarias (INIAP), Estación Experimental Santa Catalina, Quito, Ecuador

3 Embrapa Hortaliças, Brasília, Distrito Federal, Brazil

*Address all correspondence to: felipe.garces@utm.edu.ec

References

[1] Davenport LJ. Genera Solanacearum: The genera of Solanaceae illustrated, arranged according to a new system by Armando T.hunziker. Systematic Botany. 2004;**29**(1):221-222. DOI: 10.1600/036364404772974130

[2] Heiser CB, Pickersgill B. Names for the cultivated *Capsicum* species (Solanaceae). Taxon. 1969;**18**(3):277-283. DOI: 10.2307/1218828

[3] Saltos LA, Corozo-Quiñones L, Pacheco-Coello R, Santos-Ordóñez E, Monteros-Altamirano L, Garcés-Fiallos FR. Tissue specific colonization of *Phytophthora capsici* in *Capsicum* spp.: Molecular insights over plant-pathogen interaction. Phytoparasitica. 2020;**49**(1):113-122. DOI: 10.1007/s12600-020-00864-x

[4] Kumari M, Verma S, Bharat N. Effect of elevated CO^2 and temperature on incidence of diseases in bell pepper (*Capsicum annuum* L.) crop. Journal of Entomology and Zoology Studies. 2017;**6**(1):1049-1052

[5] Perry L, Flannery KV. Precolumbian use of chili peppers in the valley of Oaxaca, Mexico. Proceedings of the National Academy of Sciences. 2007;**104**:11905-11909. DOI: 10.1073/pnas.0704936104

[6] FAO. FAOSTAT: Cultivos y productos de ganadería. 2019. Recuperado 2021, de https://www.fao.org/faostat/es/#data/QCL

[7] Pinto CMF, dos Santos IC, de Araujo FF, da Silva TP. Pepper importance and growth (*Capsicum* spp.). In: Production and Breeding of Chilli Peppers (*Capsicum* spp.). Cham: Springer International Publishing; 2016. pp. 1-25

[8] Reis A, Paz-Lima ML, Moita AW, Aguiar FM, de Noronha Fonseca ME, Café-Filho AC, et al. A reappraisal of the natural and experimental host range of Neotropical *Phytophthora capsici* isolates from Solanaceae, Cucurbitaceae, Rosaceae, and Fabaceae. Journal of Plant Pathology. 2018;**100**(2):215-223. DOI: 10.1007/s42161-018-0069-z

[9] Barchenger DW, Lamour KH, Bosland PW. Challenges and strategies for breeding resistance in *Capsicum annuum* to the multifarious pathogen, *Phytophthora capsici*. Frontiers in Plant Science. 2018;**9**:628. DOI: 10.3389/fpls.2018.00628

[10] Bosland PW. Think global, breed local: specificity and complexity of *Phytophthora capsici*. In: Proceedings of the 19th International Pepper Conference, USA. 2008. pp. 14-15

[11] Kamoun S, Furzer O, Jones JD, Judelson HS, Ali GS, Dalio RJ, et al. The top 10 oomycete pathogens in molecular plant pathology. Molecular Plant Pathology. 2015;**16**:413-434. DOI: 10.1111/mpp.12190

[12] Hausbeck MK, Lamour KH. *Phytophthora capsici* on vegetable crops: Research progress and management challenges. Plant Disease. 2004;**88**:1292-1303. DOI: 10.1094/PDIS.2004.88.12.1292

[13] Kale SD, Tyler BM. Entry of oomycete and fungal effectors into plant and animal host cells. Cellular Microbiology. 2011;**13**:1839-1848. DOI: 10.1111/j.1462-5822.2011.01659.x

[14] Gaibor-Vaca DG, García-Bazurto GL, Garcés-Fiallos FR. Defense mechanisms in *Capsicum* plants against *Phytophthora*

capsici. Bionatura. 2022;**7**(1):25. DOI: 10.21931/RB/2022.07.02.25

[15] Saini SS, Sharma PP. Inheritance of resistance to fruit rot (*Phytophthora capsici* Leon.) and induction of resistance in bell pepper (*Capsicum annuum* L.). Euphytica. 1978;**27**(3):721-723. DOI: 10.1007/bf00023707

[16] Piccini C, Parrotta L, Faleri C, Romi M, Del Duca S, Cai G. Histomolecular responses in susceptible and resistant phenotypes of *Capsicum annuum* L. infected with *Phytophthora capsici*. Scientia Horticulturae. 2019;**244**:122-133. DOI: 10.1016/j.scienta.2018.09.051

[17] Bowers JH, Martin FN, Tooley PW, Luz EDMN. Genetic and morphological diversity of temperate and tropical isolates of *Phytophthora capsici*. Phytopathology. 2007;**97**(4):492-503. DOI: 10.1094/phyto-97-4-0492

[18] Granke LL, Quesada-Ocampo L, Lamour K, Hausbeck MK. Advances in research on *Phytophthora capsici* on vegetable crops in the United States. Plant Disease. 2012;**96**(11):1588-1600. DOI: 10.1094/pdis-02-12-0211-fe

[19] Rohmah B, Hadisutrisno B, Manohara D, Priyatmojo A. Karakteristik morfologi dan sebaran tipe kawin *Phytophthora capsici* asal lada di Pulau Jawa. Jurnal Fitopatologi Indonesia. 2019;**14**(5):166. DOI: 10.14692/jfi.14.5.166

[20] Smith BA, Eudoxie G, Saravanakumar D. Identification of *Phytophthora capsici* causing collar rot in hot peppers in Trinidad. Canadian Journal of Plant Pathology. 2019;**41**(1):129-134

[21] Aragaki M, Uchida JY. Morphological distinctions between *Phytophthora capsici* and *P. tropicalis* sp. nov. Mycologia. 2001;**93**(1):137-145. DOI: 10.1080/00275514.2001.12061285

[22] Rai GS, Liew ECY, Guest DI. Survey, identification and genetic diversity of *Phytophthora capsici* causing wilt of chilli (*Capsicum annuum* L.) in Bhutan. European Journal of Plant Pathology. 2020;**158**(3):655-665. DOI: 10.1007/s10658-020-02108-4

[23] Lee KJ, Kamala-Kannan S, Sub HS, Seong CK, Lee GW. Biological control of Phytophthora blight in red pepper (*Capsicum annuum* L.) using *Bacillus subtilis*. World Journal of Microbiology and Biotechnology. 2008;**24**(7):1139-1145. DOI: 10.1007/s11274-007-9585-2

[24] Lamour KH, Stam R, Jupe J, Huitema E. The oomycete broad-host-range pathogen *Phytophthora capsici*. Molecular Plant Pathology. 2011;**13**(4):329-337. DOI: 10.1111/j.1364-3703.2011.00754.x

[25] Reifschneidbr FJB, Café-Filho AC, Rego AM. Factors affecting expression of resistance in pepper (*Capsicum annuum*) to blight caused by *Phytophthora capsici* in screening trials. Plant Pathology. 1986;**35**(4):451-456. DOI: 10.1111/j.1365-3059.1986.tb02042.x

[26] Babadoost M, Pavon C, Islam SZ, Tian D. Phytophthora blight (*Phytophthora capsici*) of pepper and its management. Acta Horticulturae. 2015;**1105**:61-66. DOI: 10.17660/actahortic.2015.110

[27] Dunn AR, Smart CD. Interactions of *Phytophthora capsici* with resistant and susceptible pepper roots and stems. Phytopathology. 2015;**105**:1355-1361

[28] Steekelenburg NAM. Phytophthora root rot of sweet pepper. Netherlands Journal of Plant Pathology. 1980;**86**(5):259-264. DOI: 10.1007/bf01977301

[29] Callaghan SE, Williams AP, Burgess T, White D, Keovorlajak T, Phitsanoukane P, et al. First report of *Phytophthora capsici* in the Lao PDR. Australasian Plant Disease Notes. 2016;**11**(1):22. DOI: 10.1007/s13314-016-0210-9

[30] Ristaino JB, Johnston SA. Ecologically based approaches to management of Phytophthora blight on bell pepper. Plant Disease. 1999;**83**(12):1080-1089. DOI: 10.1094/pdis.1999.83.12.1080

[31] Hyder S, Inam-ul-Haq M, Ahmed R, Gondal AS, Fatima N, Hanan A, et al. First report of *Phytophthora capsici* infection on bell peppers (*Capsicum annuum* L.) from Punjab, Pakistan. International Journal of Phytopathology. 2018;**7**(1):51. DOI: 10.33687/phytopath.007.01.2543

[32] Naegele RP, Hausbeck MK. Evaluation of pepper fruit for resistance to *Phytophthora capsici* in a recombinant inbred line population, and the correlation with fruit shape. Plant Disease. 2014;**98**(7):885-890. DOI: 10.1094/pdis-03-13-0295-re

[33] Reis A, Café-Filho AC, Henz GP. *Phytophthora capsici*: Patógeno agressivo e comum às solanáceas e cucurbitáceas [Internet]. 2007. Available in: https://ainfo.cnptia.embrapa.br/digital/bitstream/CNPH-2009/33435/1/ct_55.pdf Embrapa.br. [retrieves in February 16, 2022]

[34] Biles CL. Relationship of Phytophthora fruit rot to fruit maturation and cuticle thickness of new Mexican-type peppers. Phytopathology. 1993;**83**(6):607. DOI: 10.1094/phyto-83-607

[35] Leonian LH. Stem and fruit blight of peppers caused by *Phytophthora capsici* sp. nov. Phytopathology. 1922;**12**:401-408

[36] Elenkov E, *Phytophthora capsici* Leon. on peppers in greenhouses. Acta Horticulturae. 1977;**58**:401-404. DOI: 10.17660/actahortic.1977.58.53

[37] Matsuoka K, Casali VMD, Saraiva TRCB. Fontes de resistência a *Phytophthora capsici* em *Capsicum annuum*. Fitopatologia Brasileira. 1984;**9**:193-201

[38] Thompson A, Uys M, Botha W. *Phytophthora capsici* (Oomycota: Fungi), a first report from South Africa. South African Journal of Botany. 1994;**60**(5):257-260. DOI: 10.1016/s0254-6299(16)30600-7

[39] Hwang BK. Phytophthora blight of pepper and its control in Korea. Plant Disease. 1995;**79**(3):221. DOI: 10.1094/pd-79-0221

[40] Moens M, Aïcha BB, Hamouda MB. Chemical control of pepper mildew *Phytophthora capsici* (Leon), on early peppers in Tunisia. Tropicultura. 1996;**4**(1):15-19

[41] Mosa AA, Zaki KI, El-Sherbeiny SN. Phytophthora root and crown rot of pepper in Egypt. Annals of Agricultural Science. 2002;**47**(3):975-991

[42] Ares JLA, Martínez AR, Pomar F, Paz JF. Telluric pathogens isolated from blighted pepper (*Capsicum annuum* L.) plants in northwestern Spain. Spanish Journal of Agricultural Research. 2005;**3**(3):326-330

[43] Camele I, Marcone C, Cristinzio G. Detection and identification of *Phytophthora* species in southern Italy by RFLP and sequence analysis of PCR-amplified nuclear ribosomal DNA. European Journal of Plant Pathology. 2005;**113**(1):1-14. DOI: 10.1007/s10658-005-8915-1

[44] Zhang ZG, Li YQ, Fan H, Wang YC, Zheng XB. Molecular detection of *Phytophthora capsici* in infected plant tissues, soil and water. Plant Pathology. 2006;**55**(6):770-775. DOI: 10.1111/j.1365-3059.2006.01442.x

[45] Sholberg P, Walker M, O'Gorman D, Jesperson G. First report of *Phytophthora capsici* on cucurbits and peppers in British Columbia. Canadian Journal of Plant Pathology. 2007;**29**(2):153-158

[46] Sheu ZM, Chen JR, Wang TC. First report of the A2 mating type of *Phytophthora capsici* infecting peppers (*Capsicum annuum*) in Taiwan. Plant Disease. 2009;**93**(5):548. DOI: 10.1094/pdis-93-5-0548c

[47] Huallanca VCA, Cadenas GCA. Control de *Phytophthora capsici* Leonian en *Capsicum annuum* cv. Papri king con fungicidas, fertilizantes y biocontroladores. Anales Científicos. 2014;**75**(1):130. DOI: 10.21704/ac.v75i1.943

[48] Benabdelkader M, Guechi A, Meacute-Zaache S. Susceptibility of Algerian pepper cultivars (*Capsicum annuum* L) to *Phytophthora capsici* strains from different geographic areas. African Journal of Biotechnology. 2015;**14**(44):3011-3018. DOI: 10.5897/ajb2015.14853

[49] Sánchez-Borges CA, Souza-Perera RA, Zúñiga-Aguilar JJ, Shrestha S, Lamour K, Castillo-Aguilar CC. First report of *Phytophthora capsici* causing damping-off of *Capsicum chinense* in the Yucatan peninsula. Plant Disease. 2016;**100**(6):1247. DOI: 10.1094/pdis-09-15-1047-pdn

[50] Nawaz K, Shahid AA, Subhani MN, Iftikhar S, Anwar W. First report of leaf spot caused by *Phytophthora capsici* on chili pepper (*Capsicum frutescens* L.) in Pakistan. Journal of Plant Pathology. 2018;**100**(1):127-127. DOI: 10.1007/s42161-018-0018-x

[51] Wartono W. Identification of the pathogen causing stem blight disease on chili in Sindangjaya Village, Cipanas, Cianjur, West Java based on morphological and molecular analyses. Jurnal AgroBiogen. 2021;**17**(1):35. DOI: 10.21082/jbio.v17n1.2021.p35-44

[52] Vélez-Olmedo JB, Saltos L, Corozo L, Bonfim BS, Vélez-Zambrano S, Arteaga F, et al. First report of *Phytophthora capsici* causing wilting and root and crown rot on *Capsicum annuum* (bell pepper) in Ecuador. Plant Disease. 2020;**104**(7):2032. DOI: 10.1094/pdis-11-19-2432-pdn

[53] Drenth A, Guest DI. *Phytophthora* in the tropics. In diversity and Management of Phytophthora in Southeast Asia. ACIAR Monograph. 2004;**114**:30-41

[54] French-Monar RD, Jones JB, Roberts PD. Characterization of *Phytophthora capsici* associated with roots of weeds on Florida vegetable farms. Plant Disease. 2006;**90**:345-350. DOI: 10.1094/pd-90-0345

[55] West P, Appiah AA, Gow NAR. Advances in research on oomycete root pathogens. Physiological and Molecular Plant Pathology. 2003;**62**(2):99-113. DOI: 10.1016/s0885-5765(03)00044-4

[56] Roberts PD, Kucharek TA. Vegetable diseases caused by *Phytophthora capsici* in Florida. UF/IFAS EDIS publication. 2018. pp. 176:1-5

[57] Jupe J, Stam R, Howden AJ, Morris JA, Zhang R, Hedley PE, et al. *Phytophthora capsici*-tomato interaction features dramatic shifts in gene expression associated with a hemi-biotrophic lifestyle. Genome Biology.

2013;**14**(6):R63. DOI: 10.1186/gb-2013-14-6-r63

[58] Fawke S, Doumane M, Schornack S. Oomycete interactions with plants: Infection strategies and resistance principles. Microbiology and Molecular Biology Reviews. 2015;**79**:263-280. DOI: 10.1128/MMBR.00010-15

[59] Siegenthaler TB, Hansen Z. Sensitivity of *Phytophthora capsici* from Tennessee to mefenoxam, fluopicolide, oxathiapiprolin, dimethomorph, mandipropamid, and cyazofamid. Plant Disease. 2021;**105**(10):3000-3007.DOI: 10.1094/PDIS-08-20-1805-RE

[60] Ristaino JB, Larkin RP, Campbell CL. Spatial and temporal dynamics of *Phytophthora* epidemics in commercial bell pepper fields. Phytopathology. 1993;**83**:1312-1320

[61] Guerrero MA, Laborde J. Current status of pepper breeding for resistance to *Phytophthora capsici* in Mexico. In: Synopses of the 4th Meeting of the *Capsicum* Working Group of Eucarpia I. V. T. Wageningen, The Netherlands; 1980. pp. 52-56

[62] Chávez-Díaz IF, Zavaleta-Mejía E. Comunicación molecular en el patosistema *Capsicum* spp. – *Phytophthora capsici*. Revista Mexicana de Fitopatología. 2019;**37**(2):251-278

[63] Lizzi Y, Roggero JP, Coulomb PJ. Behaviour of the phenolic compounds on *Capsicum annuum* leaves infected with *Phytophthora capsici*. Journal of Phytopathology. 1995;**143**(10):619-627. DOI: 10.1111/j.1439-0434.1995.tb00211.x

[64] Padilha HKM, Pereira EDS, Munhoz PC, Vizzotto M, Valgas RA, Barbieri RL. Genetic variability for synthesis of bioactive compounds in peppers (*Capsicum annuum*) from Brazil. Food Science and Technology. 2015;**35**(3):516-523. DOI: 10.1590/1678-457x.6740

[65] Low PS, Merida JR. The oxidative burst in plant defense: Function and signal transduction. Physiologia Plantarum. 1996;**96**(3):533-542. DOI: 10.1111/j.1399-3054.1996.tb00469.x

[66] Stoessl A, Unwin CH, Ward EWB. Postinfectional inhibitors from plants. I. Capsidiol, an antifungal compotmd from *Capsicum frutescens*. Phytopathology. 1972;**Z. 74**:141-152

[67] Egea C, Alcazar MD, Candela ME. Capsidiol: Its role in the resistance of *Capsicum annuum* to *Phytophthora capsici*. Physiologia Plantarum. 1996;**98**(4):737-742. DOI: 10.1111/j.1399-3054.1996.tb06679.x

[68] Alabouvette C, Olivain C, Steinberg C. Biological control of plant diseases: The European situation. European Journal of Plant Pathology. 2006;**114**(3):329-341. DOI: 10.1007/s10658-005-0233-0

[69] Pal KK, McSpadden GB. Biological control of plant pathogens. The Plant Health Instructor. 2006; p. 1-25. DOI: 10.1094/PHI-A-2006-1117-02

[70] Bhusal B, Mmbaga M. Biological control of Phytophthora blight and growth promotion in sweet pepper by *bacillus* species. Biological Control. 2020;**150**:104373. DOI: 10.1016/j.biocontrol.2020.104373

[71] Ita MNVD, Fátima JH, Lezama CP, Simón AB, Cortés GL, Romero-Arenas O. Bio-controller effect of four native strains of *Trichoderma* spp., on *Phytophthora*

capsici in Manzano chili (*Capsicum pubescens*) in Puebla-Mexico. Journal of pure and applied. Microbiology. 2021;**15**(2):998-1005. DOI: 10.22207/jpam.15.2.58

[72] Mmbaga MT, Gurung S, Maheshwari A. Screening of plant endophytes as biological control agents against root rot pathogens of pepper (*Capsicum annum* L.). Journal of Plant Pathology & Microbiology. 2018;**9**:435. DOI: 10.4172/2157-7471.1000435

[73] Garcés-Fiallos FR, Saltos A, Corozo-Quiñonez L, Pacheco-Coello L, Santos-Ordóñez R, Urresta LF, et al. *Capsicum* hypocotyls mycobiome diversity is unaffected by *Phytophthora capsici* inoculation. Physiological and Molecular Plant Pathology. 2022;**118**:101801. DOI: 10.1016/j.pmpp.2022.101801

[74] Narayanasamy P. Biological disease management systems for agricultural crops. In: Biological Management of Diseases of Crops. Progress in Biological Control. Vol 16. Springer, Dordrecht. 2013. pp. 189-235. DOI: 10.1007/978-94-007-6377-7_6

[75] Alejo NL, Guzmán-Plazola RA, Mejía EZ, Rincón VHA, Escobar VA. Etiology and evaluation of control alternatives for wilt in Chile de arbol (*Capsicum annuum* L.) in La Vega, Metzitlán, Hidalgo, México. Revista Mexicana de Fitopatología. 2015;**3**(1):31-50

[76] Parra G, Ristaino JB. Resistance to mefenoxam and metalaxyl among field isolates of *Phytophthora capsici* causing Phytophthora blight of pepper. Plant Disease. 2001;**85**:1069-1075. DOI: 10.1094/PDIS.2001.85.10.1069

[77] Dunn AR, Milgroom MG, Meitz JC, McLeod A, Fry WE, McGrath MT, et al. Population structure and resistance to mefenoxam of *Phytophthora capsici* in New York state. Plant Disease. 2010;**94**:1461-1468. DOI: 10.1094/PDIS-03-10-0221

[78] Wang W, Liu D, Zhuo X, Wang Y, Song Z, Chen F, et al. The RPA190-pc gene participates in the regulation of metalaxyl sensitivity, pathogenicity and growth in *Phytophthora capsici*. Gene. 2021;**764**:145081. DOI: 10.1016/j.gene.2020.145081

[79] FRAC. Fungal Control Agents Sorted by Cross Resistance Pattern and Mode of Action [Internet]. 2020. Available from: https://www.frac.info/ [Accessed: July 2, 2021]

[80] Castro A, Flores J, Aguirre M, Fernández SP, Rodríguez G, Osuma P. Traditional and molecular studies of the plant pathogen *Phytophthora capsici*: A review. Journal of Plant Pathology & Microbiology. 2014;**5**:245-253. DOI: 10.4172/2157-7471.1000245

[81] Saltos LA, Monteros-Altamirano A, Reis A, Garcés-Fiallos FR. *Phytophthora capsici*: The diseases it causes and management strategies to produce healthier vegetable crops. Horticultura Brasileira. 2022;**40**(1):005-017. DOI: 10.1590/s0102-0536-20220101

[82] Gilardi G, Baudino M, Moizio M, Pugliese M, Garibaldi A, Gullino M. Integrated management of *Phytophthora capsici* on bell pepper by combining grafting and compost treatment. Crop Protection. 2013;**53**:13-19. DOI: 10.1016/j.cropro.2013.06.008

[83] Razdan VK, Gupta S, In: Peshin R, Dhawan AK. Editors.

Integrated Pest Management: Innovation-Development Process. Jammu: Springer; 2009. pp. 369-389. DOI: 10.1007/978-1-4020-8992-3_15

[84] Sanogo S, Ji P. Integrated management of *Phytophthora capsici* on solanaceous and cucurbitaceous crops: Current status, gaps in knowledge and research needs. Canadian Journal of Plant Pathology. 2012;**34**(4):479-492. DOI: 10.1080/07060661.2012.732117

Chapter 5

Major Pests and Pest Management Strategies in the Sweet Pepper (*Capsicum annuum*)

Aman Dekebo

Abstract

Sweet peppers (*Capsicum annuum*) (Solanaceae) fruits have been used as a food ingredient in Peru for more than 8,000 years. Then gradually, the plant has been cultivated in several countries worldwide. The fruits of the plant can be added to soups and stews as spices. These were reported to treat fevers, seasickness, muscle sprains, or soreness. Thrips, whiteflies, mites, and aphids were critical pests in sweet peppers. Therefore, effectively managing this important fruit to improve its yields and quality is very important. Pesticides have harmful effects on the environment and health of people. Therefore, alternative pest management strategies become more advisable to control pests of sweet pepper. These strategies including intercropping of sweet pepper with other plants, oviposition deterrents, natural enemy release, use of resistant cultivars, and eliciting plant defenses are implemented as environment-friendly control methods.

Keywords: sweet peppers, *capsicum*, intercropping, pests, management strategies

1. Introduction

Sweet pepper (*C. annuum* L.) and tomato crops occupy most of the area among protected species around the world. Pepper cultivation is almost entirely carried out in open fields; however, the extension of protected, greenhouse-cultivated peppers has intensively increased [1]. Sweet pepper is one of the very important crops worldwide. *C. annuum* includes hot as well as sweet pepper varieties. The world's main pepper-producing countries include China, Mexico, and Turkey, with over 17.4, 2.7, and 2.5 million pepper tons in 2016, respectively [2].

Other European countries and Canada are important producers of greenhouse peppers, with 135 million kg of peppers grown where yield is ≤12 t/ha, and sweet pepper is subject to several pests [3], such as beetles [4], caterpillars [5], aphids [6], and thrips [7]. Alternative control methods of pests to exclusively use of insecticides and integrated pest management (IPM) such as cultural, biological, and chemical treatments were used to manage sweet pepper pests [3].

2. Major pests of sweet pepper in green house

Sweet pepper is susceptible to attacks by several pests, which reduces fruit quality and yield. When insects and mites attack sweet pepper, both direct and indirect damages occur [3, 8]. The indirect damage is caused by pests when they transmit viruses, while the direct damage is those occurring when pests themselves cause damage to different parts of the plants. Instead of controlling pests with insecticides, integrated pest management strategies, especially those based on biological control methods, have been used effectively for several years worldwide to control pests of sweet pepper. Sweet pepper is one of the highly attractive crops to pests and pathogens. Those pests affecting pepper crops can differ based on geographic area and cropping system such as open field or greenhouse, and conventional or organic farming.

2.1 Arthropod pests

2.1.1 Thrips

Thrips are the number one pest of greenhouse crops in different climatic regions mainly due to their polyphagous diet and their ability to rapidly develop resistance to commercially available insecticides. Thrips cause significant damage, such as feeding and ovipositing on pepper leaves, fruit, and flowers, leading to decreased quality and marketability of fruits. The most known damaging thrip species include the western flower thrips, *Frankliniella occidentalis*, *Thrips tabaci* Lindeman, and *Thrips palmi* Karny [9]. Another thrip attacking *Capsicum* species is *Scirtothrips dorsalis* Hood. Thrips usually like closed areas such as the flowers, under the calyx of fruit and young leaves, which make them difficult to control with insecticides [3].

2.1.2 Whiteflies

Two principal whiteflies are reported to attack sweet pepper: the greenhouse whitefly, *Trialeurodes vaporariorum* (Westwood), and *Bemisia tabaci*. In some European regions, such as Spain, *B. tabaci* is the major pest of peppers [10].

Adult whiteflies and nymphal whiteflies feed on the vascular tissue in plants (phloem), causing direct damage. On the other hand, indirect damage results from virus transmission by adults and sooty mold, which develops on their excreted honeydew [11, 12].

2.1.3 Mites

2.1.3.1 Spider mites

The two-spotted spider mite (*Tetranychus urticae*, Koch) and the carmine spider mite (*T. cinnabarinus*, Boisduval) were reported as the important pests of pepper worldwide. These mites cause whitish or yellowish stippling in the upper leaf surfaces and produce silk webbing [13].

2.1.3.2 Broad mites

The broad mite is one of the notorious pepper pests in different regions of the world [14]. It mainly causes damage to different parts of sweet pepper at the younger stage and is difficult to manage due to its small size. It usually feeds on the lower leaf surface and distorts flowers and blistering of fruits. When the broad mite gains access to enter a greenhouse, it can spread rapidly, resulting in high economic losses [3].

2.1.4 Aphids

Aphids are generally important pests of sweet pepper, especially in open fields compared with those covered. The most critical aphid is the green peach aphid, *Myzus persicae* (Sulzer), which causes direct damage to leaves by the secretion of phytotoxins into the plants [15]. In general, the damage caused by aphids is indirect, such as sooty molds growing on secreted honeydew and transmission of potyviruses. Another aphid pest reported in subtropical-covered peppers is *Aphis gossypii* (Glover) [16].

3. Pest management

Management strategies of greenhouse pests using chemicals were reported to result in several problems, such developments of resistance to chemicals by pests, and environmental and health problems are caused by those chemical pesticides [17]. The possibility of applying biological control programs to problems of greenhouse pests is highly recommended. Even though they will not completely solve the problems, they can reduce pest populations to an acceptable level. Biological control generally requires more time than pesticides to bring a pest population under an acceptable control level [18]. Biological control strategy such as releasing different predators and parasitoids was reported to be environmentally friendly production method of sweet pepper. Applying the biological control program resulted in a high yield of sweet pepper production (35.06%) compared with the control [17]. In this book chapter, various pest control methods are reviewed.

3.1 Polyethylene plastic cladding material

The most commonly used material for greenhouse covering is polyethylene screening plastic, which has an ultraviolet-absorbing characteristic. The material was prepared by adding a specific UV-absorbing compound to raw polyethylene, which makes the plastic material that can stop the transmission of 95% of the UV light (200–380 nm) and transmits 80% of visible light (380–700 nm) [19]. The behaviors of insects can be affected by the modification of UV light. Light at 360–400 nm activates whiteflies (*T. vaporariorum*) to walk and fly [20]. Similarly, the effects of UV-absorbing plastics on thrips, *F. occidentalis*, have also been reported as host-finding behavior is disrupted [21]. Reduction in reproduction in aphids by deactivation of its flying behavior was reported. For example, reproduction in *M. persicae* is significantly reduced in a greenhouse covered with UV-absorbent plastic [15].

3.2 Predators

3.2.1 Phytoseiidae (Acari)

Many phytoseiid mites are predatory, and several species have been developed as biological control strategies [3]. McMurtry and Croft [22] classify several phytoseiid mites based on their feeding behaviors. Type I species include *Phytoseiulus* spp., mites that are well known as predators of webbing spider mites. Type II mite species include *Neoseiulus californicus* (McGregor) and *N. fallacis* (Garman), which feed spider mites and others. Type III species, for example, *N. barkeri* (Hughes), *N. cucumeris* (Oudemans), and *Iphiseius degenerans* (Berlese), are generalists that often prefer prey other than spider mites (in whose webbing they may become entangled) and thrips [3].

3.2.2 Diptera

Aphidoletes aphidimyza is a predatory cecidomyiid that has been used to control aphids [23, 24]. *A. aphidimyza* in the adult stage are important for hunting aphids. They are released in the greenhouse together with the parasitoid *A. colemani* Viereck for a better effect. Additionally, their larvae were reported to attack several species of aphids. A method of effective utilization in pepper fields in greenhouses was highlighted [25].

3.2.3 Heteroptera

Weintraub [3] reviewed many species of the anthocorid bug, *Orius*, which have been evaluated for controlling thrips on protected sweet pepper crops. Both adults and nymphs can frequently be found in flowers, a location that is favorable for thrips. Additionally, unlike predatory mites, *Orius* spp. attack both adult and immature thrips. Dissevelt *et al*. [26] reported the potential of *O. niger*, *O. majusculu*, and *O. laevigatu*, *O. insidiosus* and *O. albidipennis* to control thrips on pepper.

3.2.4 Neuroptera

Chrysoperla carnea (Stephans) is a generalist predator that appears in open fields and is difficult to rear commercially because it is highly carnivorous. There was a report [27] on controlling the aphid *M. persicae* on peppers by *C. carnea*.

3.2.5 Mirid predator

Predatory mirid bugs (Hemiptera: Miridae) were reported as biocontrol agents in sweet pepper [28, 29]. In addition to their agricultural uses as predators, mirid predators can induce plant defenses by phytophagy [30]. The punctures caused by mirid plant feeding induced the release of a mixture of volatile organic compounds (VOCs), namely green leaf volatiles [((*Z*)-3-hexenyl acetate, (*Z*)-3-hexenyl propanoate, (*Z*)-3-hexenyl butanoate, (*Z*)-3-hexenyl 3-methylbutanoate, and (*Z*)-3-hexenyl benzoate) and their common precursor (*Z*)-3-hexenol], methyl salicylate, and octyl acetate. Octyl acetate was detected in *M. pygmaeus*-punctured plants, which repelled the herbivore pests *F. occidentalis* and *B. tabaci* and simultaneously attracted the whitefly parasitoid *Encarsia Formosa* [30].

3.2.6 Aphidophagous hoverflies

Moerkens et al. [31] investigated the potential of hoverflies *Eupeodes corollae* and *Sphaerophoria rueppellii* to manage foxglove aphid *Aulacorthum solani* in sweet pepper. In a semi-field study, aphid numbers were significantly lower in the *E. corollae* and *S. rueppellii* treatments than in control. The fruit yield and seed set were also increased for *E. corollae* and *S. rueppellii*.

3.3 Parasitoids

Most parasitoids developed for biological control belong to the family Aphelinidae. These parasitoids are tiny wasps and have been used to control the whitefly [32]. For instance, *Encarsia formosa* is a parasitoid used worldwide for the biological control of whiteflies attacking various vegetables such as sweet pepper cultivated in greenhouses. It lays eggs into hosts and causes a reduction of hosts. Uses of *E. formosa* are well known for controlling whiteflies because of the following factors. Whitefly population growth is reduced when *E. formosa*'s intrinsic rate of increase is greater than the host's intrinsic in the presence of parasitoids. This situation has resulted when host plants facilitate parasitoid searching and exhibit partial resistance to whitefly development. Additionally, giving-up time on infested leaves increases when hosts or host products are located, increasing the likelihood that parasitoids will encounter suitable hosts in a patch. It was also reported that spatial refuges for whiteflies from parasitoids exist in large greenhouses (greater than 1000 m^2) and consequently promote stable host or parasitoid dynamics [32].

3.4 Entomopathogens

Entomopathogenic fungi are effective methods of controlling pests though it requires humid conditions to allow the propagules to penetrate the insect body, after which fungal development generally proceeds. A means of pathogen delivery is by using bees that were reported to be applicable for field crops [33, 34]. For instance, strains of *Beauveria bassiana* (Balsamo) Vuillemin have been shown to be virulent to *F. occidentalis* and *T. tabaci* [35, 36]. Bumble bees were used for pepper pollination and could distribute the *B. bassiana* conidia to flowers and leaves [37]. Since *N. cucumeris* is known to be unaffected by *B. bassiana*, [36], they have been reported to deliver inoculum for days or weeks without compromising other biological control methods.

3.5 Oviposition deterrents

Luteolin 7-O-β-D-apiofuranosyl-(1 2)-β-D-glucopyranoside was isolated from matured leaves of sweet pepper and identified as the ovipositional deterrent against *Liriomyza trifolii* (Burgess), the American serpentine leaf miner [38]. This insect species attacks *C. annuum* leaves in the young stage. The attack by this insect to leaves of the plant decreases as the plant becomes matured. This compound completely deterred *L. trifolii* females from laying their eggs on a host plant leaf treated at 4.90 g/cm^2 [38]. Additionally, phytol [(2*E*)-3,7,11,15-tetramethyl-2-hexadecen-1-ol] constituent of matured leaves of sweet pepper was also reported as an ovipositional deterrent against *L. trifolii*. Phytol [(2*E*)-3,7,11,15-tetramethyl-2-hexadecen-1-ol] completely deterred the females from laying their eggs on host plant leaves treated at 35.2 μg/cm^2. 4-Aminobutanoic acid, (2S,4R)-4-hydroxy-1-methyl-2-pyrrolidine carboxylic acid, and 4-amino-1--D-ribofuranosyl-2(1H)-pyrimidinone were reported from the leaves of sweet pepper showed oviposition deterrence toward adult flies of *L. trifolii* from laying their eggs on kidney bean leaves (host plant) treated at 3.70, 16.60, and 6.45 g/cm^2, respectively [39].

3.6 Intercropping of sweet pepper with other plants

It was reported that monocropped pepper, such as pepper, intercropped with maize (*Zea mays*) or eggplant (*Solanum melongena*). Maize acted as a barrier crop for *aphids* (*Aphis gossypii*) and reduced virus infection on pepper in the first part of the cropping season [40, 41]. Eggplant was reported to act as a trap crop for aphids and reduced virus infection on pepper for a longer period than maize [40].

Additionally, Li et al. [42] reported intercropping rosemary with sweet pepper results in the population of three main pest species on sweet pepper. The significant pest population suppression and the absence of adverse effects on natural enemies in the sweet pepper/rosemary intercropping system show the potential of this strategy in the IPM framework. Consequently, intercropping sweet pepper increases yields of pepper intercropped with other crops.

3.7 Eliciting plant defenses in sweet pepper

The exposure of sweet pepper plants to HIPVs such as [*(Z)*-3-hexenol, *(Z)*-3-hexenyl acetate, *(Z)*-3-hexenyl propanoate, *(Z)*-3-hexenyl butanoate, hexyl butanoate, methyl salicylate, and methyl jasmonate] over 48 h activates the sweet pepper immune defense system against various pests [43]. The volatiles inducted the plant defenses due to the regulation of the jasmonic acid and salicylic acid signaling pathway. A principal sweet pepper pest is *Frankliniella occidentalis*, and *Orius laevigatus* is its main natural enemy. HIPV-exposed sweet pepper plants were investigated, and the results showed that only plants exposed to *(Z)*-3-hexenyl propanoate and methyl salicylate repelled *F. occidentalis,* whereas *O. laevigatus* showed a strong preference for plants exposed to *(Z)*-3-hexenol, *(Z)*-3-hexenyl propanoate, *(Z)*-3-hexenyl butanoate, methyl salicylate, and methyl jasmonate. In the related study, treatment of cotyledons of sweet pepper with 50 and 100 μM of jasmonic acid (JA) solution resulted in strong plant's oviposition deterrence against the leaf miner *L. trifolii* [44]. These results demonstrate that HIPVs act as elicitors to sweet pepper plant defenses by enhancing defensive signaling pathways. Enhancing defensive signaling pathways is very important for integrating HIPVs-based approaches in sweet pepper pest management systems, which may provide a sustainable strategy to manage insect pests in horticultural plants.

3.8 Resistant cultivars

Several pepper accessions have been evaluated for thrips resistance, and significant differences in damage levels have been observed [45]. The difference between different accessions was found to be through tolerance. There were attempts to breed pepper plants for simultaneous resistance to arthropod vectors and pathogens though those attempts were not successful. Some pepper accessions were reported to be resistant to *A. gossypii*; however, these plants did not show resistance to the green peach aphid attack. Several commercial peppers exhibited strong resistance to *M. persicae*, as demonstrated by reduced damage to the plants [46]. Interestingly, some pepper cultivars exhibited tolerance to aphid-transmitted viruses [3].

3.9 Chemical control

Weintraub [3] reviewed that even insecticides considered acceptable for use along with some beneficial organisms. For instance, Spinosad that is prepared by fermentation of an actinomycete has been evaluated to control Western flower thrips (WFT) [47]. It was found that while Spinosad was effective against immature and adult WFT, it also showed low toxicity to *Amblyseius cucumeris* exposed to leaves 1 day after treatment. Spinosad exhibited moderate toxicity to *Orius insidiosus* 1 and 8 days after treatment and high toxicity to *Encarsia formosa* up to 28 days after application [47].

4. Conclusion

The major pests of sweet pepper are Arthropod pests such as thrips, whiteflies, mites, and aphids. These pests cause enormous damage on sweet pepper by feeding and ovipositing on pepper leaves, fruit, and flowers, which lead to decreased quality and marketability of fruits. Some of the various pest management strategies used to manage these pests are physical technique (polyethylene plastic cladding materials), biological control methods such as the use of predators, parasitoids, and entomopathogens, oviposition deterrents, intercropping with other plants, eliciting plant defenses-resistant cultivars, and chemical control. Utilizing a biological control strategy by releasing different predators and parasitoids resulted in an environmentally friendly method of controlling sweet pepper greenhouse production.

Acknowledgements

Adama Science and Technology University supported the author with the grant, ASTU/AS-R/003/2020. I thank the World Academy of Sciences (TWAS) and the United Nations Educational, Scientific, and Cultural Organization (UNESCO) for funds allocated to the author under the TWAS Research Grant RGA No. 20-274 RG/CHE/AF/AC_G – FR3240314163.

Conflict of interest

The author declares that they have no conflicts of interest.

Author details

Aman Dekebo[1,2,*]

1 Department of Applied Chemistry, Adama Science and Technology University, Adama, Ethiopia

2 Institute of Pharmaceutical Sciences, Adama Science and Technology University, Adama, Ethiopia

*Address all correspondence to: amandekeb@gmail.com

References

[1] Bouagga S. Enhancing pest management in sweet pepper by the exploitation of zoophytophagy. (Ph. D. Thesis). Universitat Jaume I. Castelló de la Plana. 2018

[2] Faostat F. FAOSTAT Statistical Database. Rome, Italy: FAO (Food and Agriculture Organization of the United Nations); 2016

[3] Weintraub PG. Integrated control of pests in tropical and subtropical sweet pepper production. Pest Management Science: Formerly Pesticide Science. 2007;**63**:753-760

[4] Eller FJ, Bartelt RJ, Shasha BS, Schuster DJ, Riley DG, Stansly PA, et al. Aggregation pheromone for the pepper weevil, Anthonomus eugenii Cano (Coleoptera: Curculionidae): Identification and field activity. Journal of Chemical Ecology. 1994;**20**:1537-1555

[5] Ruberson JR, Herzog GA, Lambert WR, Lewis WJ. Management of the beet armyworm (Lepidoptera: Noctuidae) in cotton: Role of natural enemies. Florida Entomologica. 1994:77:440-453

[6] da Cunha LC, Resende RO, Nagata T, Inoue-Nagata AK. Distinct features of Pepper yellow mosaic virus isolates from tomato and sweet pepper. Fitopatologia Brasileira. 2004;**29**:663-667

[7] Funderburk J, Stavisky J, Olson S. Predation of Frankliniella occidentalis (Thysanoptera: Thripidae) in field peppers by Orius insidiosus (Hemiptera: Anthocoridae). Environmental Entomology. 2000;**29**:376-382

[8] Gullino ML, Albajes R, Nicot PC. Integrated Pest and Disease Management in Greenhouse Crops. Vol. 9. Swizerland: Springer; 2020

[9] Lewis T. Thrips as Crop Pests. Wallingford, UK: Cab International; 1997:1-13

[10] Stansly PA, Calvo J, Urbaneja A. Release rates for control of Bemisia tabaci (Homoptera: Aleyrodidae) biotype "Q" with *Eretmocerus mundus* (Hymenoptera: Aphelinidae) in greenhouse tomato and pepper. Biological Control. 2005;**35**:124-133

[11] Gorman K, Hewitt F, Denholm I, Devine GJ. New developments in insecticide resistance in the glasshouse whitefly (*Trialeurodes vaporariorum*) and the two-spotted spider mite (*Tetranychus urticae*) in the UK. Pest Management Science: Formerly Pesticide Science. 2002;**58**:123-130

[12] Elbert A, Nauen R. Resistance of Bemisia tabaci (Homoptera: Aleyrodidae) to insecticides in southern Spain with special reference to neonicotinoids. Pest Management Science. 2000;**56**:60-64

[13] Zhang Z-Q. Mites of Greenhouses: Identification, Biology and Control. Wallingford, UK: Cabi; 2003

[14] Gerson U. Biology and control of the broad mite, Polyphagotarsonemus latus (Banks) (Acari: Tarsonemidae). Experimental & Applied Acarology. 1992;**13**:163-178

[15] Chyzik R, Dobrinin S, Antignus Y. Effect of a UV-deficient environment on the biology and flight activity ofMyzus persicae and its hymenopterous parasite Aphidius matricariae. Phytoparasitica. 2003;**31**:467-477

[16] Castané C. Status of biological and integrated control in greenhouse vegetables in Spain: Successes and challenges. IOBC WPRS Bulletin. 2002;**25**:49-52

[17] El Arnaouty S, El-Heneidy A, Afifi AI, Heikal I, Kortam MN. Comparative study between biological and chemical control programs of certain sweet pepper pests in greenhouses. Journal of Biology Pesticide Control. 2020;**30**:1-7

[18] Kortam M.. Biological Control of Certain Greenhouse Pests. CU Theses. 2019

[19] Antignus Y, Ben-Yakir D. Ultravioletabsorbing barriers, an efficient integrated pest management tool to protect greenhouses from insects and virus diseases. In: Insect Pest Management. Berlin, Heidelberg: Springer; 2004

[20] Coombe P. Visual behaviour of the greenhouse whitefly, Trialeurodes vaporariorum. Physiological Entomology. 1982;**7**:243-251

[21] Vernon R, Gillespie D. Spectral responsiveness of Frankliniella occidentalis (Thysanoptera: Thripidae) determined by trap catches in greenhouses. Environmental Entomology. 1990;**19**:1229-1241

[22] McMurtry J, Croft B. Life-styles of phytoseiid mites and their roles in biological control. Annual Review of Entomology. 1997;**42**:291-321

[23] Adams R Jr, Prokopy R. Aphidoletes aphidimyza (Rondani) (Diptera: Cecidomyiidae): An effective predator of the apple aphid (Homoptera: Aphididae) in MassachusettsProt. Ecology. 1980;**2**:27-39

[24] Meadow R, Kelly W, Shelton A. Evaluation ofAphidoletes aphidimyza [Dip.: Cecidomyiidae] for control ofMyzus persicae [Hom.: Aphididae] in greenhouse and field experiments in the United States. Entomophaga. 1985;**30**:385-392

[25] van Schelt J, Mulder S. Improved methods of testing and release of Aphidoletes aphidimyza (Diptera: Cecidomyiidae) for aphid control in glasshouses. European Journal of Entomology. 2000;**97**:511-516

[26] Dissevelt M, Altena K, Ravensberg W. Comparison of different Orius species for control of Frankliniella occidentalis in glasshouse vegetable crops in the Netherlands. In: Comparison of Different Orius Species for Control of Frankliniella Occidentalis in Glasshouse Vegetable Crops in the Netherlands. 1995. pp. 839-845

[27] El-Arnaouty S, Gaber N, Tawfik M. Biological control of the green peach aphid Myzus persicae by Chrysoperla carnea (Stephens) sensu lato (Neuroptera: Chrysopidae) on green pepper in greenhouses in Egypt. Pest Control. 2000;**10**:109-116

[28] Messelink GJ, Bloemhard CM, Kok L, Janssen A. Generalist predatory bugs control aphids in sweet pepper. IOBC/wprs Bulletin. 2011;**68**:115-118

[29] Messelink G, Bloemhard C, Hoogerbrugge H, Van Schelt J, Ingegno BL, Tavella L. Evaluation of mirid predatory bugs and release strategy for aphid control in sweet pepper. Journal of Applied Entomology. 2015;**139**:333-341

[30] Bouagga S, Urbaneja A, Rambla JL, Flors V, Granell A, Jaques JA, et al. Zoophytophagous mirids provide pest control by inducing direct defences, antixenosis and attraction to parasitoids in sweet pepper plants. Pest Management Science. 2018;**74**:1286-1296

[31] Moerkens R, Boonen S, Wäckers FL, Pekas A. Aphidophagous hoverflies reduce foxglove aphid infestations and improve seed set and fruit yield in sweet pepper. Pest Management Science. 2021;77:2690-2696

[32] Hoddle M, Van Driesche R, Sanderson J. Biology and use of the whitefly parasitoid Encarsia formosa. Annual Review of Entomology. 1998;**43**:645-669

[33] Gross HR, Hamm JJ, Carpenter JE. Design and application of a hive-mounted device that uses honey bees (Hymenoptera: Apidae) to disseminate Heliothis nuclear polyhedrosis virus. Environmental Entomology. 1994;**23**:492-501

[34] Butt T, Carreck N, Ibrahim L, Williams I. Honey-bee-mediated infection of pollen beetle (Meligethes aeneus Fab.) by the insect-pathogenic fungus, Metarhizium anisopliae. Biocontrol Science Technology. 1998;**8**:533-538

[35] Gindin G, Barash I, Raccah B, Singer S, Ben-Ze'ev I, Klein M. The potential of some entomopathogenic fungi as biocontrol agents against the onion thrips, Thrips tabaci and the western flower thrips, Frankliniella occidentalis. Folia Entomologica Hungarica. 1996;**57**:37-42

[36] Jacobson R, Chandler D, Fenlon J, Russell K. Compatibility of beauveria bassiana (balsamo) vuillemin with amblyseius cucumeris oudemans (acarina: Phytoseiidae) to control frankliniella occidentalis pergande (thysanoptera: Thripidae) on cucumber plants. Biocontrol Science and Technology. 2001;**11**:391-400

[37] Shipp L, Broadbent B, Kevan P. Biological control of Lygus lineolaris (Hemiptera: Miridae) and Frankliniella occidentalis (Thysanoptera: Thripidae) by Bombus impatiens (Hymenoptera: Apidae) vectored Beauveria bassiana in greenhouse sweet pepper. Biological Control. 2006;**37**:89-97

[38] Kashiwagi T, Horibata Y, Mekuria DB, Tebayashi S-I, Kim C-S. Ovipositional deterrent in the sweet pepper, Capsicum annuum, at the mature stage against Liriomyza trifolii (Burgess). The Biochemist. 2005;**69**:1831-1835

[39] Dekebo A, Kashiwagi T, S-i T, Kim C-S. Nitrogenous ovipositional deterrents in the leaves of sweet pepper (Capsicum annuum) at the mature stage against the leafminer, Liriomyza trifolii (Burgess). The Biochemist. 2007;**71**:421-426

[40] Hussein MY, Abdul Samad N. Intercropping chilli with maize or brinjal to suppress populations of *Aphis gossypii* Glov., and transmission of chilli viruses. International Journal of Pest Management. 1993;**39**:216-222

[41] Kahn BA. Intercropping for field production of peppers. Hort Technology. 2010;**20**(3):530-532

[42] X-w L, Lu X-x, Z-j Z, Huang J, J-m Z, Wang L-k, et al. Intercropping rosemary (*Rosmarinus officinalis*) with sweet pepper (*Capsicum annum*) reduces major pest population densities without impacting natural enemy populations. Insects. 2021;**12**:74

[43] Riahi C, González-Rodríguez J, Alonso-Valiente M, Urbaneja A, Pérez-Hedo M. Eliciting plant defenses through herbivore-induced plant volatiles' exposure in sweet peppers. Frontiers in Ecology and Evolution. 2022;**9**:776827. DOI: 10.3389/fevo.2021.776827

[44] Tebayashi S, Horibata Y, Mikagi E, Kashiwagi T, Mekuria DB, Dekebo A,

et al. Induction of resistance against the leafminer, Liriomyza trifolii, by jasmonic acid in sweet pepper. Bioscience. 2007;**71**:70033-1-6

[45] Fery RL, Schalk JM. Resistance in pepper (*Capsicum annuum* L.) to western flower thrips [Frankliniella occidentalis (Pergande)]. Horticultural Science. 1991;**26**:1073-1074

[46] Frantz JD, Gardner J, Hoffmann MP, Jahn MM. Greenhouse screening of Capsicum accessions for resistance to green peach aphid (*Myzus persicae*). Horticultural Science. 2004;**39**:1332-1335

[47] Jones T, Scott-Dupree C, Harris R, Shipp L, Harris B. The efficacy of spinosad against the western flower thrips, Frankliniella occidentalis, and its impact on associated biological control agents on greenhouse cucumbers in southern Ontario. Pest Management Science. 2005;**61**:179-185

Chapter 6

Capsicum: Breeding Prospects and Perspectives for Higher Productivity

Raman Selvakumar,
Dalasanuru Chandregowda Manjunathagowda
and Praveen Kumar Singh

Abstract

Chipotle peppers were grown in America before being carried to Europe by Columbus. Capsicum breeding began with choosing wild species for desired characteristics, with additional development based on precision selection. To improve capsicum yields, traditional methods such as mass selection, pedigree, single-seed descent, backcrossing, and hybridization are being used. Capsicum has a high level of genetic diversity due to multiple new gene rearrangements. Capsicum fruits are high in nutrients that are beneficial to human health. As a result, the world market for and consumption of capsicum has lately grown. Capsicum breeding programmes aim to improve yield, biotic, abiotic resistance, and nutritional quality. Recent breakthroughs in capsicum breeding have included introgression, mutation breeding, polyploidy, haploidy, embryo rescue, and the use of genetic markers. Molecular technology has grown into an important tool that, when coupled with classic selection and hybridization procedures, has the potential to result in great success in an established capsicum genetic breeding programme.

Keywords: pepper, molecular, embryo, quality, haploidy

1. Introduction

Capsicum is cultivated all over the world for a variety of applications needing different levels of quality and characteristics. Fresh or dried capsicums have been considered a foodstuff for many years due to the fact that they contain all of the required nutrients. Capsicum fruits have double the amount of vitamin C found in citrus fruits. In contrast, dried red chilies are high in vitamin A and β-carotene, making them a healthy snack option [1]. According to photochemical studies, it also has cancer-prevention due to the presence of higher antioxidant properties [2]. Capsaicin lotion is used therapeutically to treat arthritis and other painful chronic conditions [3]. Capsicum extracts are used in the production of cosmetics and pharmaceuticals. Cultivating capsicum in pots or gardens as an ornamental plant is becoming trendier these days [4]. Sweet pepper is the most important spice traded across the globe. It is possible to classify the capsicum market into five broad divisions depending on the fruit's shape and intended purpose. In the fresh market, whole fruits are available in green or red; in the fresh-processing market, sauce, paste, canning, and pickles are available; in the dried spice market, whole fruits and pepper powder are available; and in the decorative market, ornamental varieties are available [5]. The ultimate objective of capsicum breeding projects will change according to the needs of farmers and consumers. Biotic and abiotic stresses are taken into account during the breeding of capsicum varieties and hybrids.

2. Breeding history

Capsicums are believed to have originated in the Western Hemisphere and have been used as a food source since 7500 BC. They are said to have originated in South America and spread to Central America. Capsicum was brought to Europe by Christopher Columbus, and it quickly spread across Africa and Asia. The genetic inheritance of important agro-horticultural traits, mutant forms, male sterility, disease, pest resistance, and quality characteristics were all required for the purpose of early capsicum study. It has been said in a number of places that these characteristics are driven by single genes with a dominant or recessive mode of action, and that some of these characteristics are governed by quantitative trait loci. *Capsicum annuum* is by far the most significant member of the *Capsicum* family, owing to the fact that it is the species of the genus *Capsicum* that is most often economically produced. When plant breeding first started, it was based mostly on the ability of individuals to be selected, which was a time-consuming and unstructured process. Bringing the notions of Mendelian genetics and inheritance into the field of vegetable breeding was a pivotal moment in the history of the field. Currently, plant breeders are using a variety of ways to select for desirable characteristics in their crops, with the methodology chosen depending on the objectives of the breeding programme [6, 7]. The primary goals of capsicum breeders are genetic enhancement in productivity, biotic and abiotic resistance and nutraceutical compounds.

2.1 Contemporary objectives of Capsicum breeding

Depending on the location, the breeding goals for capsicum, both hot and bell pepper, varies depending on the nation of culture, the purpose of cultivation, the growing conditions, the end user, and the preferences of the customers. Some countries prefer peppers that are fiery and pungent, while others prefer peppers that are sweet. The diseases that harm the crop also differ depending on the climate that prevails in the different countries. Biotic and abiotic resistance breeding, on the other hand, is one of the most important goals in the development of capsicum varieties [8]. A comprehensive depiction of the numerous pest and diseases that affect capsicum has been presented by Pohronezny [9]. The process of disease resistance breeding begins with the discovery of resistant sources, followed by a study of their genetics, and finally with the introduction of promising genotypes. It has been shown that in the case of capsicum, a significant amount of disease and pest resistance from wild species has been introduced into commercial cultivars in order to increase disease resistance. It is also necessary to assess the amount of crossability across species when developing an interspecific hybridization programme for resistance gene introgression. The utilization of wild materials for the insertion of biotic resistance genes into desirable cultivars has produced significant contributions to crop improvement, most notably in terms of increased yield and quality, as well as stability in capsicum production, among other applications. Introgression efforts to introduce disease resistance genes into superior cultivars have frequently failed when disease resistance traits are under polygenic control and linked with undesirable horticultural and economic characteristics. To counteract the ongoing growth and emergence of new disease races and strains against presently existing resistant genotypes, it is vital to seek for and deploy new resistant sources on a regular basis.

The second goal for which capsicum breeders across the globe are trying is to increase yield, which will ultimately result in increased total production. In this regard, the heterosis breeding programme is becoming more important. It is preferred that more emphasis be placed on the development of F_1 hybrids based on available male sterility systems, since this reduces the amount of time and work necessary for hybrid seed production. In order to make hybrids, both genetic (GMS) and cytoplasmic male sterility (CMS) systems have been used, with the cytoplasmic male sterility system being the most extensively used. The discovery of new CMS sources, the identification of their maintainers, and the diversity of CMS systems all become key goals as a result of this process. The discovery of restorers with excellent general and specific combining capacity, as well as the insertion of resistance genes into these CMS lines and restorers, should also be a priority for the generation of hybrids.

Capsicum breeding aims are also influenced by market demand and end-use usability, among other factors. This involves breeding for horticultural economic and nutritional quality traits, among other things. Fresh market breeders are looking for characteristics such as fruit color at the unripe stage, which is often green (light, medium, or dark), fruit size-length, width, and pericarp thickness. Furthermore, the amount of pungency is an important and distinct feature of capsicum breeding that should not be overlooked. Understanding consumer preferences for pungency in a given location is an extremely important aspects of the research process. Pungency, which is a major economic characteristic of capsicum, is attributable to the presence of a chemical complex of alkaloids known as capsaicinoids in the plant [10]. Capsaicin and dihydrocapsaicin are the two major capsaicinoids in capsicum, accounting for around 90% of the total capsaicinoids in most pungent varieties.

One of the most important quality attributes that capsicum breeders consider when developing commercial varieties is the amount of capsaicin present in the plant [11]. Capsanthin concentration in capsicum is estimated using the high-performance liquid chromatography (HPLC) analytical method. The capsanthin-capsorubin synthase (CCS) enzyme is found only in the *Capsicum* genus, and it is responsible for the production of two red pigments, capsanthin and capsorubin [12]. According to the USDA, red capsicum is being bred for higher capsanthin content in order to be used as a dried spice and for industrial extracts viz, paprika oleoresin, capsaicinoids, and carotenoids. The red color of chilli peppers is indicative of the presence of capsorubin and capsanthin, whereas the yellow color of chilli peppers is due to the presence of β-carotene and violaxanthin [13]. In general, the higher the ASTA color rating, the deeper the genotype's ripening red color. It has a capsanthin content of between 70 and 100 ASTA units (low), 71–100 ASTA units (medium), and 101–150 ASTA units (high). The ASTA color affects the brightness of a product, while the surface color affects the hue. As a consequence, another major aim of capsicum breeding is the development of paprika varieties that meet the great demand for nonpungent pods with a high color value for industrial uses. Dry matter content is a critical trait to breed for in dry capsicum, since it is used to make dry powder and whole dried fruits. Additionally, these are the most often requested characters for export reasons. While a high dry-matter content in red chilli fruit is useful commercially, there is no correlation between the dry-matter content and the fruit's capsaicin concentration [14]. It is critical to have a thin pericarp for dry capsicum in order to facilitate drying. The surface of fruits with a thick pericarp gets wrinkled and their appearance becomes bland as a consequence of drying. In response to increasing industrialization, the risk of crop failure associated with climate change, and consumer demand (both domestic and international) for more nutritious and safer foods, increased emphasis is being placed on breeding genotypes with increased tolerance to high temperatures, drought, and wider adaptability.

2.2 Major goals of Capsicum breeding

A variety of colors, including medium or dark green at the unripe stage and red, yellow, or orange at the mature stage, are among the key targets of genetic enhancement of sweet peppers. Research in this initiative aims to find and develop new varieties of capsicum that are rich sources of antioxidants as well as vitamins. Flavonoids and carotenoids (red, yellow, and orange carotenoids), which contain vitamin A precursors such as alpha and beta carotene, as well as β-cryptoxanthin are also included in this category [15, 16]. Breeding efforts are also focused at increasing fruit set and yield in varying climatic situations, including open and protected. Low temperatures, drought, and salt stress are all being studied as part of breeding efforts to combat abiotic threats. Breeding for long-term storage stability of carotenoid extract and resistance to *Phytophthora* fruit rot are also on the research agenda. Powdery mildew, anthracnose, *Phytophthora* fruit rot, bacterial wilt and viruses, are some of the most common diseases that affect sweet peppers in open-field as well as in the green house production. Another important objective is to develop sweet pepper genotypes that are more tolerant to tropicalization, since this is a cool-season crop. Produce will be accessible throughout the year for a longer amount of time in places that are not usual [17]. The primary breeding goals in protected culture are to develop sweet pepper lines with an indeterminate growth habit, tolerant to training and pruning systems, blocky fruit, resistance to biotic stress, and resistance to root-knot nematode [18].

3. Breeding strategies

3.1 Conventional breeding approaches

Capsicum crop improvement has been achieved by the applying of conventional breeding procedures such as mass selection, pureline selection, pedigree breeding, single-seed descent method, backcross breeding, and heterosis breeding. Different types of breeding strategies, including mutation breeding and polyploidy breeding, have also been used in an effort to generate variety in capsicum, which may then be used in improvement initiatives. At the time of the beginning of systematic plant breeding, many tactics for capsicum improvement were used: mass selection, pureline selection, pedigree breeding, single-seed descent method, and backcross breeding were among those employed. Mass selection is one of the most straightforward strategies that has been utilized to increase the quality of capsicum. Improvement for many qualities with simple inheritance may be done at the same time without having to worry about the pedigree of the individuals involved. Initially, it was employed to enhance landraces or open-pollinated cultivars of capsicum, which were previously unproductive. Characters with high heritabilities may be readily repaired using this technique, but an acceptable amount of variability is also retained. Traditional landraces and local cultivars were the primary targets of pure line selection since farmers were cultivating them. This strategy involves selecting better plants, harvesting them individually, then evaluating their progenies the following year in order to determine plant performance. Progenies that exhibit better performance and are free of genetic variability are collected in large quantities and assessed further in repetitive experiments against check cultivar(s). So, this approach has been widely utilized to develop various types of capsicum for commercial cultivation, and it is still being used today.

In a pedigree selection method, selection is conducted between and within families of individuals, and selected individuals are issued a pedigree number, allowing any offspring in any generation to be traced back to its original crop that was picked in the F_2 generation. This has historically been one of the most often utilized methods for developing capsicum cultivars. Selecting superior parental cultivars is critical for the development of this approach. It is often employed in conjunction with backcrossing to effectively introduce essential genes into advanced inbreds. Using the single-seed descent (SSD) procedure, one seed from a single fruit is collected from each plant in a segregating generation. The segregating generation is produced in greenhouse conditions to increase the number of generations each year. Additionally, this enables the formation of a large number of pure inbred lines for use in test crossings for the development of hybrids, as well as the generation of recombinant inbred line populations for mapping research. In the capsicum breeding programme, the backcross method is the most often used way for disease resistance development. This method is most often used to transfer a single gene or a limited number of genes from primitive cultivars or wild forms to leading cultivars. In exceptional circumstances, even BC_2 families may be routed using the pedigree strategy (modified backcross) rather than the usual backcrossing process, which involves 5–6 backcrosses with the recurrent parent. While open-pollinated varieties of hot peppers and bell peppers are still commonly available, heterosis breeding has been found to increase hot pepper and bell pepper production. Numerous hybrids have been developed in the capsicum plant; nevertheless, the hybrid research effort should be continued to ensure that seeds are affordable to farmers. Capsicum F_1 hybrids are gaining popularity as a consequence of a large number of private sector seed companies investing in vegetable industry research and seed manufacturing. Male sterility is frequently used in the generation of hybrid seeds in the chilli plant to increase the cost-effectiveness of seed production. The discovery of various male-sterile mutants, which eliminate the need for more laborious emasculation techniques, together with the identification of several marker genes, has improved the detection of undesirable types at the seedling stage even further. GMS and cytoplasmic-genetic male sterility (CGMS) are two types of genetic male sterility that are now being economically exploited in chilli for hybridization. GMS has been suggested above CGMS for hybrid seed production because GMS exhibits male sterility and male fertility segregation, but CGMS does not.

CGMS was discovered in capsicum for the first time by Peterson [19] and was designated as USDA accession PI164835. There have been no reports of any additional CMS sources so far. "*orf507*" and "*tp6-2*" are two aberrant mitochondrial genes found in the capsicum CMS system that have been linked to male sterility [20]. Because the genes are present in the mitochondria, they are passed down via the maternal line. It is also necessary that a nuclear gene for the restoration of fertility be absent in order for male sterility to be expressed. A restorer line is required for effective hybrid seed development when the restoration of fertility is driven by a single dominant gene, as is the case in most cases. In order to preserve male sterility, a maintainer line must include both fertile cytoplasm and the lack of a nuclear gene that would allow for fertility restoration. Due to the fact that the CGMS system of hybrid seed production necessitates the use of three lines, namely, the CMS line, the keeper of male-sterile line, and a restorer of fertility in hybrids, the system is referred to as the three-line system of hybrid seed production in the capsicum plant. The GMS technique has also been employed to develop capsicum, but to a lesser degree. In the GMS system, the expression of male sterility is regulated by homozygous recessive genes (*ms*/*ms*), while male fertility is controlled by homozygous dominant or heterozygous genes (*Ms*/*MS* or *Ms*/*ms*). *Ms*/*Ms* and *Ms*/*ms* are isogenic lines that vary solely at the *Ms* locus, and they are essential for the maintenance of male sterility in the GMS population. Intercrossing between these two lines results in offspring that are a combination of both male fertile (*Mf*/*ms*) and male sterile (*ms*/*ms*) sperm in equal quantities. Visual identification identifies male fertile plants in the field and rejects them, while male-sterile lines are utilized for hybrid seed production [21]. To begin with, the production of successful capsicum cultivars was largely dependent on the breeder's expertise, perception, and good fortune in selecting promising genotypes. Cultivar development still relies on the breeder's knowledge and perception, even in the age of cutting-edge breeding procedures. A blend of both science and art, plant breeding continues to be such. There are several more approaches for improving capsicum, including mutation breeding, polyploid and haploid development, transgenics, and marker aided breeding, all of which have had some success.

3.1.1 Colchiploidy breeding

After the newly produced polyploid has grown in strength, it will next adapt to its new environmental surroundings. It has been suggested that the advantage of polyploids over diploids might be due to the phenomena of transgressive segregation, which is the production of extreme phenotypes, as described by Van de Peer and co-workers [22]. It has been suggested by Malhova [23] that capsicum may react to variations in ploidy in the same manner as *Solanum* does. Capsicum ploidy levels may be intentionally increased or decreased in a very straightforward manner. Using colchicine to repair injured leaf axils, it is possible to achieve somatic doubling in plants. On the other side, synthetic autotetraploids do not seem to have any economic or breeding benefits over diploids. Polyploid capsicum is characterized by slowed growth and the presence of larger, thicker, and dark green leaves [24]. The presence of more chloroplasts and larger chloroplasts in polyploid leaves has been attributed to the polyploid leaves' rich green tint [25]. When compared to diploid capsicum, the tetraploid capsicum has increased leaf, stem, and root dry weight, as well as increased leaf area and thickness. Tetraploids have been shown to have an increased capacity for water, NO_3–N, and K absorption, which correlates with an increase in photosynthesizing potential; they also generate small but more unified fruits irrespective of fruit loading; and they produce tinier but more uniform-sized fruits regardless of fruit loading [26].

It has been discovered that the tetraploid capsicum flowers about one month later than the diploids. The total number of flowers produced was reduced, with this reduction owing mostly to the non-branching character of the polyploidy [24]; nonetheless, the total number of flowers produced was increased. Raghuvanshi and Sheila [25] discovered that the colchiploids of *Capsicum frutescens* had delayed and protracted blooming, as well as a bigger and more diversified number of floral components than the diploids. Polyploids have larger blooms as well as larger pollen grains, which are also typical of polyploids [27].

Treatment of seeds with colchicine resulted in the generation of tetraploid plants of the *C. annuum* variety "Chigusa," thus according to Ishikawa et al. [28]. Following a flow cytometric study of the seeds treated with colchicine, it was observed that 20% of the seeds were tetraploid. Tetraploid flowers had seven petals and filaments, 20 of ovaries, and 25% larger pollen grains than diploid flowers, which typically had six petals and anthers [29]. Tetraploid blooms were also 20% bigger than diploid blooms in diameter. Polyploids have also been shown to be sterile, which might be owing to abnormalities seen during the meiotic phase [30]. Following treatment with colchicine, researchers discovered that a plant of the chilli pepper *cv.* CO-2 had chromosomal counts ranging from 2n = 38 to 96. It possessed 4.95% pollen fertility but did not produce any seeds, and its development was inhibited as a result [31]. Although colchicines have been used to double the number of homozygotes produced by anther culture, these homozygotes have not yet been exploited to produce commercial F_1 hybrids capable of exhibiting heterosis. Instead, they have been used to investigate the genetic mechanisms of resistance to pests [32] and diseases [33].

Malhova [23] successfully established an interspecific hybrid of *Capsicum pubescens* and *Capsicum annuum* by fertilizing *C. pubescens* with autotetraploid *Capsicum annuum* pollen. The use of induced auto-tetraploidy to overcome post-fertilization hurdles may benefit future interspecific crosses of the capsicum genus, according to this research. Pochard [34, 35] revealed previously unknown trisomies in *C. annuum*. Pochard [34] demonstrated that trisomies may be used to identify genes on specific chromosomes, either via skewed segregation ratios seen in the offspring of trisomic F_1 hybrids or through dosage effects shown when trisomics are compared to conventional diploid humans [36]. These trisomics confirmed the presence of gene "*C*" (which determines pungency) on acrocentric chromosome number "XI" [34] and its location on the long arm [37], since the trait pungency segregated independently of the markers on the acrocentric chromosomes' short arms [34].

3.1.2 Embryo rescue

Embryo rescue has been the most often used technique to overcome post-zygotic hybridization difficulties in interspecific crosses. The hybridization of capsicum species from separate gene pools has been observed, but incompatibility has also been reported within the same gene pool, such as between *C. annuum* and *Capsicum chinensis* or between *C. annuum* and *Capsicum frutescens*. Many fruits with shriveled seeds are created as a result of incorrect endosperm and/or embryo formation in numerous interspecific crosses in the *Capsicum* spp. As a result, many fruits with shriveled seeds are produced, which are unable to germinate properly. In the scientific literature, it has been reported that a hybrid embryo originating from interspecific crosses in the *Capsicum* genus has been successfully recovered. Fari et al. [38] performed the first successful attempt at embryo rescue in the capsicum genus, acquiring an embryo from a hybrid between *C. annuum* and *C. baccatum*. This was the first successful attempt at embryo rescue in the *Capsicum* genus. It is yet another example of extensive hybridization in which immature interspecific embryos or embryos from different species were/were rescued prior to abortion, as was the case with the hybridization of *C. annuum* and *C. baccatum* [39].

Technically, the procedures of embryo removal and in vitro embryo culture are both highly challenging to perform. The stage at which embryo abortion occurs during hybridization may also be determined by the genotypes of the individuals engaged in the cross, according to some researchers. Yoon et al. [40] reports that some researchers have been successful in saving interspecific embryos at the most advanced stages of development in the *Capsicum* genus, while other researchers have had to save them at the very beginning of development [40, 41]. Rescue of embryos at an earlier stage, on the other hand, is more difficult, and the possibility of recovering interspecific hybrids is lower at this time [42]. Anthracnose resistance exhibited in *C. baccatum* lines has been shown to have been transferred into *C. annuum* lines by the rescue of embryos obtained from interspecific crossings between the two species and subsequent culture of embryos resulting from those crosses [40].

Alternate methods of overcoming the aforementioned issue may be used, such as the construction of a genetic bridge based on the usage of species that are phylogenetically closer to the two species that are affected by crossability barriers. A bridge species must be used in conjunction with this method because it must be capable of crossing with both the target species and the target species' predator. Initially, it is essential to cross the bridge species with one of the target species, and then to cross the resultant hybrid with the other target species [39]. After doing this research, it was shown that *C. chinensis* is an acceptable bridge species for performing widespread hybridization between the species *C. annuum* and *C. baccatum* [43].

3.2 Modern breeding approaches

3.2.1 Development of Capsicum Haploids

Capsicum annuum and *Capsicum frutescens* anther cultures were used to produce the first haploids in the genus *Capsicum* [44], which were then used to produce the first haploids in the genus *Capsicum* [45]. Because of the poor recovery of haploid plants from androgenic cultures seen in previous studies, researchers decided to construct experiments with the goal of identifying the elements that influence the induction of androgenesis. Based on the various experiments conducted on haploid induction, the androgenic response was determined to depend on growing conditions, age, the genotype of the donor plant [46], and developmental stage of microspores in the anther.

The development of doubled haploids is one of the most effective means of establishing full homozygosity in any crop species; nevertheless, because of the plant's recalcitrance, its application in capsicum enhancement is still restricted [47]. Capsicum breeding requires a genetically stable and homozygous plant population in order to better understand genetics, as well as mapping and identification of genes for various morphological traits and biotic and abiotic stress-related morphotypes. Despite the limited frequency of findings, a number of researches on the practical side of haploid breeding in several capsicum species are now underway [48]. It has previously been reported that parental lines created utilizing doubled haploid (DH) technology may be used to create varieties and F_1 hybrids [49]. DH capsicum lines also had higher production attributes and dry matter content in their fruits [50]. Superior DH lines with great variation in plant and fruit features, as well as androgenic capsicum lines with favorable qualities, have been isolated [51]. In addition to enhanced production, it has been feasible to develop Capsicum DHs with improved quality characteristics such as fruit shape, flavor, fruit hardness, dry matter content, total soluble content, phenolic content, and antioxidant activity, such as CUPRAC and FRAP [52, 53].

Nowaczyk et al. [54] used DH technology to improve the shelf life of soft-flesh *Capsicum* spp. recombinants. DH lines derived from in vitro capsicum anther culture showed varying degrees of resistance to *Xanthomonas campestris* pv. *vesicatoria* [55] and *Phytophthora capsici* [55] and *Phytophthora capsici* [56]. These disease-resistant DH lines might be exploited to develop novel genotypes that are resistant to many diseases. Using anther culture, it was also possible to get PVY-resistant lines as well as lines with important qualitative and quantitative traits [57]. Todorova et al. [58] revealed their findings after using haploid culture to generate capsicum lines with high production, enhanced fruit attributes, and decreased sensitivity to *Verticillium* wilt. Microspore embryogenesis has been used to create genotypes with higher productivity, resistance to *Verticillium dahliae* Kleb [48] and resistance to tobacco mosaic virus [59].

3.2.2 Techniques of genetic modifications

Genetic transformation has been proposed as an alternative method for the enhancement of capsicum. Transgenic technology, in the case of capsicum, has many key benefits, one of which is that it allows for the transfer of valuable genes or the acquisition of distinctive features across interspecific and intergeneric barriers. A pioneering research on the transformation of capsicum was initially published in 1990 [60]. The lack of repeatability in the pepper plant, on the other hand, is a significant stumbling hurdle for capsicum transformation studies. There has been a great deal of work done on capsicum transformation for disease resistance, particularly against viruses such as tobacco mosaic virus (TMV), pepper mild mottle virus (PMMV) [61], tomato mosaic virus (ToMV) [62], cucumber mosaic virus (CMV) [63]. A transgenic virus resistance strategy that makes use of viral coat protein regions and satellite RNA is referred to as RNA silencing in the scientific community [64]. The use of transformation and overexpression of *TsiI*, a tobacco pathogenesis-related (PR) gene in capsicum, allowed us to demonstrate broad spectrum resistance against a variety of pathogens, including PMMV and CMV, as well as the bacteria *Xanthomonas campestris* pv. *vesicatoria* and the fungal pathogen *Porphyromonas capsici* [62]. A limit is placed on the number of transformation experiments in capsicum that are carried out on parameters other than disease resistance.

Harpster and colleagues discovered that the enzyme ripening-related endo-1.4-b-glucanase is inhibited in transgenic capsicum [65]. Transformation research in capsicum that includes the introduction of foreign genes from other plants or species are quite usual. It was possible to develop a dwarf transgenic capsicum after transformation with the *OsMADS1* gene from rice [66, 67]. Ketoacyl-ACP reductase (*CaKR1*) was discovered by using RNA silencing to identify a unique gene in the capsicum plant that produces non-pungent fruits, which was named by the researchers [68]. A reporter gene for capsicum transformation studies has been most often used, and it is the *GUS* gene (β-glucuronidase) [61]. It has been most common to create capsicum transgenics using *Agrobacterium*-mediated transformation, with cotyledons and/or hypocotyls being used as explants in the vast majority of experiments [61]. Capsicum transgenic *C. frutescens* has recently been exposed to direct transformation by the gene gun, resulting in the development of a new variety of capsicum [69].

3.2.3 Marker-assisted breeding

In the field of capsicum enhancement, marker-assisted breeding (MAB), also called molecular-assist breeding, has gained favor. Capsicum has been studied using isozyme markers, amplified fragment length polymorphism (AFLPs), random amplified polymorphic DNA (RAPD), restriction fragment length polymorphism (RFLPs), simple sequence repeat (SSR), single nucleotide polymorphism (SNP), and COS II markers. These markers have been widely used to study the transmission of important features, as well as to identify horticultural and disease resistance genes, as well as quantitative trait loci (QTLs).

3.2.3.1 Necessity of molecular breeding

However, environmental factors make it difficult to select for quantitatively passed-down complex characteristics, making it difficult to use direct selection on genotype or phenotypic values. As a result, the indirect selection is seen to be a preferable method of selection. Conventional breeding has had little success because of the polygenic regulation of resistance characteristics, the large variety of pathogen strains found in various habitats, the complexity of the host-pathogen relationship, and the great diversity in pathogenicity. There are several reasons for this. It's possible to use indirect selection by looking at other, more readily measurable features that are closely connected to the desired traits, but which are more difficult to measure or which are impacted by the environment.

Due to other features, indirect selection for yield is constrained. The inability to select for certain genes is typically due to a lack of available tools, facilities, and resources. As a result of the development of molecular (DNA) markers, plant breeders now have an effective tool for doing gene selection. Marker-assisted gene selection is not a true form of gene selection, but it is the best method available for indirectly selecting target genes in DNA. A reliable and successful method is the marker-assisted selection (MAS). Both Collard and Mackill [70] and Kole and Gupta [71] have shown the benefits of MAS over traditional phenotypic selection. Compared to phenotypic breeding, selection utilizing molecular markers is easier.

Selecting a single plant with high dependability may be done at any step of the plant's life cycle, in addition to this. Gene localization and the generation of novel genotype combinations with high yield and stress-resistant genes have both been made possible by the advent of molecular markers. This speeds up the breeding process considerably. They've helped researchers learn more about how certain genes work. Gene placement and selection aren't the only things that molecular markers help with; they may also be used to analyze genetic diversity, monitor quality, and aid in breeding. The use of molecular markers is critical to expediting the speed of improvement programmes in order to fulfill the rising demand for increased capsicum yield and disease-resistant genotypes. Capsicum molecular markers [72] have been used for DNA fingerprinting, genetic diversity analysis, QTL analysis of important biotic stresses, and MAS [73].

Capsicum genotypes may be reliably differentiated by estimating their genetic diversity. Different kinds of marker systems such as isozymes, RAPD [74], AFLP [75], and SSR [76] have been used for genetic diversity study and varietal identification in capsicum. The use of molecular markers to determine genetic diversity is helpful for a variety of reasons, including choosing different parent combinations for hybrid production, understanding the evolutionary link of various *Capsicum* species, and accurately identifying varietals. In order to protect and make use of plant genetic resources, it is necessary to do molecular characterization on germplasm. Conventional plant breeding has certain drawbacks, which MAS attempts to alleviate using molecular selection. Numerous genetic markers in capsicum, including mapped microsatellites and single nucleotide polymorphisms (SNPs), have been utilized successfully in genomics [77].

SSRs and SNPs have been used to clone and define genes in capsicum that influence stress tolerance, quality traits, and other aspects of plant growth. These genes are valuable assets for molecular-assisted breeding. Capsicum researchers now employ SSRs as the most common markers, in part because of their widespread availability in the public domain, as well as their ease of use and efficacy [78]. Genomes/QTLs for a wide range of important traits in capsicum, such as pungency, fertility restoration, soft flesh and deciduous fruits [79], capsanthin content, fruit size and shape [80], male sterility [81], parthenocarpy [82], resistance to CMV [83], potyvirus, have been identified in the genomes of various species of capsicum.

3.2.4 TILLING and Eco-TILLING approaches

Genetic differences are created via mutations, which are the most common cause of genetic diversity. It is currently considered to be a cornerstone of contemporary plant breeding. In the case of capsicum, it has been discovered that mutation breeding is a successful and efficient breeding strategy. Daskalov [84] has provided an in-depth analysis of this topic matter. The seeds of capsicum are the most appealing portions to be treated with mutagenic agents. It is advisable to utilize seeds of uniform size and germinability (96–100%), as well as seeds with a low moisture content (approximately 13%), in order to achieve high repeatability of findings. Ionizing radiation utilized as a mutagen should result in a 40–60% chance of survival [85], but chemical mutagens should result in a 70–80% chance of survival [86]. Bell peppers, as opposed to spicy peppers, are often more radiosensitive in general. Pollen grains have also been treated with gamma or X-rays and utilized for the pollination of emasculated, non-irradiated flowers immediately after irradiation. In order to prevent cross-pollination, the M_1 generation (first generation after mutagen treatment) plants must be cultivated on separate plots (at least 700 m away from other capsicum plants) followed by bagging of the M_1 flowers to prevent out-crossing. Each experiment requires the cultivation of at least 3000–5000 M_1 plants. Plants are produced in the following generation at a rate of 20–25 M_2 plants per M_1 plant or 10–15 M_2 plants per M_1 fruit (with 2–3 fruits per M_1 plant) in the first generation. According to the M_2 field population size estimates, the population size ranges between 70,000 and 100,000 plants, however this varies depending on the kind of selection to be done and the number of observations to be made. The M_2 generation is the one in which the majority of the work is done in terms of mutant selection. All identified mutants must be selfed, which is commonly accomplished by bagging the flowers, to enable offspring testing.

In capsicum, the method of mutation breeding has been employed extensively for functional gene annotation as well as for the creation of new variability that can be exploited in breeding [87] used the sweet pepper cultivar "*Maor*" to develop a mutation population that was later used for the isolation and characterization of genes controlling plant architecture and flowering [87]. Similar mutant populations have been established in chilli peppers using the cultivar "*Yuwol-cho*," which is a cross between two different cultivars [88]. TILLING (targeted induced local lesions in genome) technique was used in the same cultivar "*Yuwol-cho*" by Jeong et al. [89], and they were successful in isolating a line that was resistant to the tobacco etch virus (TEV).

Capsicum mutant populations were generated by Daskalov [90, 91] by the use of X-rays and gamma irradiation, and these populations were studied further. Novel male-sterile lines were isolated from these populations and then described to determine their suitability for use in breeding programmes. These populations were also used to generate capsicum cultivars that have desirable features like resistance to the cucumber mosaic virus (CMV), superior taste, greater yield, and compact plant height, among others [90, 91]. Japanese researchers Honda and colleagues [92] generated mutants using heavy ion beams (12C and 20Ne), however the majority of the screening was done in the M_1 generation, which is the first generation of mutants. Capsicum has been subjected to ultraviolet irradiation in order to produce mutants with higher levels of vitamin C and E [93]. Three male-sterile lines were obtained from a capsicum mutant population produced by gamma irradiation and used in hybrid development by Daskalov and Mihailov [93].

Tomlekova and colleagues [15, 16] have recovered mutants with altered shoot architecture in hot pepper [86], some induced mutants in sweet pepper [92], and capsicum with enhanced β-carotene and orange color on maturity [15, 16]. *Capsicum annuum* L. dry seeds were gamma irradiated, and numerous intriguing mutants were developed, the most interesting of which were induced male-sterile mutations, which were acquired after gamma irradiation of the seeds. Male sterility is controlled by a small number of recessive genes, which are designated as *ms-3*, *ms-4*, *ms-6*, *ms-7*, and *ms-8*. Using the male-sterile lines Pazardjishka kapia *ms-3* and Zlaten medal *ms-8* that were recovered following mutagen treatment, the researchers were able to evaluate their combining capacity against the original male-sterile line that was utilized for hybridization. The results obtained suggest that there is no statistically significant difference in the combining ability for early and total yields, according to the findings. Several male-sterile lines were crossed with a huge number of other lines in order to produce hybrid combinations that might be used for a variety of reasons. When it came to early yield, the majority of the hybrid combinations outperformed the check. There was also a rise in overall yield in several hybrids, which was observed. Two-hybrid combinations, designated "*Krichimski run*" and "*Lyulin*" were released as cultivars based on the male-sterile lines retrieved from the mutant population and used in the development of the hybrid combinations.

4. Limitations of capsicum breeding

Crop improvement via conventional plant breeding relies on manipulating plant genomes inside the core gene pool of a genus. Hybridization and selection are used in conjunction with backcross breeding, mutagenesis, and somatic hybridization to develop novel combinations of genomes from various species. Segregating progeny phenotypic evaluations are used to identify economically significant novel characteristics. Beyond a certain point, traditional plant breeding's relevance in improving quality and output becomes exceedingly challenging. Modern breeding distinguishes itself from traditional breeding by separating phenotypes from genotypes. The phenotype is a manifestation of a person's inherited genes in a particular environment. Genotype selection and screening are focused on phenotypic expression rather than genetic variation. As a consequence, new cultivars include qualities that breeders want, but they also have undesirable characteristics that were not taken into account during the selection process, and this transfer of undesirable traits from existing to new varieties is almost always unavoidable via traditional breeding. Breeders face a second problem when they attempt to make use of the genetic variation that is present in groups that are incompatible with each other. Wide-scale hybridization and intensive backcrossing of created hybrids with recipient parents provide novel features into cultivated types. Nevertheless, the targeted features of interest don't appear on their own; they're accompanied by larger portions of wild chromosomes and so are linked to linkage drag, which may include unwanted genes. Traditional breeding cannot regulate the expression of target genes in a new genetic background, which is the third constraint. Using current breeding methods like marker-assisted selection helps speed up the introgression process while also reducing linkage drag. Plant breeding will continue to use conventional techniques to create new and better varieties, in other words. In contrast to traditional phenotypic selection, molecular breeding plays an important role since it is more accurate, quick, and cost-effective than conventional phenotypic selection.

5. Future perspectives

The success of Capsicum breeding demonstrates the program's potential for future growth. Capsicum has a lot of opportunities for development and expansion. The genetic diversity of capsicum has been discovered using a whole-genome sequence and a genotyping by sequencing technique to locate single nucleotide polymorphisms (SNPs). Because of this genomic information, we may now believe that the genetic makeup of capsicum can be changed to a far greater extent than previously anticipated. The number of studies that relate genetic variation to observable phenotypic variation, on the other hand, is still fairly low. Finding unique linkages between the generated genetic resources and crucial capsicum characteristics such as fruit size, production, pungency, resilience to abiotic stress, nutritional content, and disease resistance is a major research subject. Furthermore, the utilization of transgenic technology in capsicum is slow because of the difficulty of changing and regenerating capsicum. Now that the capsicum genome sequence is accessible, researchers may look at the most recent genome-editing technologies and their potential use in genetic upgrading of capsicum. The capacity of organisms to profit from gene/genome editing is greatly limited by a lack of well-characterized target gene information. Combining cutting-edge genetic breeding technologies with tried-and-true processes like as traditional selections and crosses will be important in capsicum breeding in the future.

Conflict of interest

The authors declare no conflict of interest.

Author details

Raman Selvakumar[1], Dalasanuru Chandregowda Manjunathagowda[2*]
and Praveen Kumar Singh[1]

1 ICAR-Indian Agricultural Research Institute, New Delhi, India

2 ICAR-Directorate of Onion and Garlic Research, Pune, India

*Address all correspondence to: dcmgowda@gmail.com

References

[1] Shetty AA, Magadum S, Managanvi K. Vegetables as sources of antioxidants. Journal of Food and Nutritional Disorders. 2013;**2**(1):1-5

[2] Pramanick K, Srivastava SK. Role of capsaicin in cancer prevention. In: Srivastava SK, editor. Role of Capsaicin in Oxidative Stress and Cancer. Netherlands: Springer; 2013. pp. 1-18

[3] Bhattacharya A, Chattopadhyay A, Mazumdar D, Chakravarty A, Pal S. Antioxidant constituents and enzyme activities in chilli peppers. International Journal of Vegetation Science. 2010, 2010;**16**:201-211

[4] Bosland PW, Votava EJ. Peppers: Vegetable and Spice Capsicums. 2nd ed. London, UK: CABI Publishing; 2012

[5] Poulos JM. Pepper breeding (*Capsicum* spp.): Achievements, challenges and possibilities. Plant Breeding Abstracts. 1994;**64**:143-155

[6] Singh P, Cheema DS, Dhaliwal MS, Garg N. Heterosis and combining ability for earliness, plant growth, yield and fruit attributes in hot pepper (*Capsicum annuum* L.) involving genetic and cytoplasmicgenetic male sterile lines. Scientia Horticulturae. 2014;**168**:175-188

[7] Singh R, Giri SK, Kotwaliwale N. Shelf-life enhancement of green bell pepper (*Capsicum annuum* L.) under active modified atmosphere storage. Food Packaging and Shelf Life. 2014;**1**:101

[8] Paran I, Vander Voort JR, Lefebvre V, Jahn M, Landry L, van Schriek M, et al. An integrated genetic linkage map of pepper (*Capsicum* spp.). Molecular Breeding. 2004;**13**:251-261

[9] Pohronezny K. Compendium of Pepper Diseases. Minnesota, USA: American Phytopathological Society; 2003. p. 63

[10] Perucka I, Materska M. Phenylalanine ammonialyase and antioxidant activities of lipophilic fraction of fresh pepper fruits *Capsicum annuum* L. Innovative Food Science & Emerging Technologies. 2001;**2**:189-192

[11] Hachiya S, Kawabata F, Ohnuki K, Inoue N, Yoneda H, Yazawa S, et al. Effects of CH-19 Sweet, a non-pungent cultivar of red pepper, on sympathetic nervous activity, body temperature, heart rate, and blood pressure in humans. Bioscience, Biotechnology, and Biochemistry. 2007;**71**:671-676

[12] Guzman I, Hamby S, Romero J, Bosland PW, Connell MAO. Variability of carotenoid biosynthesis in orange colored *Capsicum* spp. Plant Science. 2011;**179**(1-2):49-59

[13] Englewood CNJ. Pungency of Capsicums and Their Oleoresins (HPLC Method). New York: American Spice Trade Association; 1985. pp. 111-114

[14] Dhall RK. Breeding for quality traits in chilli: A review. Journal of Research: Punjab Agricultural University. 2008;**45**(3 & 4):156-160

[15] Tomlekova N, Timina OO, Timin OY. Achievement and perspectives of sweet pepper breeding towards high beta carotene. Acta Horticulturae. 2009;**1**:205-209

[16] Tomlekova NB, Timina OO, Timin OY. Achievements and perspectives of sweet pepper breeding

towards high beta-carotene. Acta Horticulturae. 2009;**830**:205-212

[17] Ferrara A, Lovelli S, Di Tommaso T, Perniola M. Flowering, growth and fruit setting in greenhouse bell pepper under water stress. Journal of Agronomy. 2011;**10**:12-19

[18] de Swart EAM. Potential for breeding sweet pepper adapted to cooler growing conditions—A physiological and genetic analysis of growth traits in Capsicum [Ph.D. thesis]. Wageningen, The Netherlands: Wageningen University

[19] Peterson PA. Cytoplasmically inherited male sterility in *Capsicum*. The American Naturalist. 1958;**92**:111-119

[20] Gulyas GK, Pakozdi JS, Lee Y. Analysis of fertility restoration by using cytoplasmic male-sterile red pepper (*Capsicum annuum* L.) lines. Breeding Science. 2006;**56**:331-334

[21] Shifriss C. Male sterility in pepper (*Capsicum annuum* L.). Euphytica. 1997;**93**(1):83-88

[22] Van de Peer Y, Fawcett JA, Proost S, Sterck L, Vandepoele K. The flowering world: A tale of duplications. Trends in Plant Science. 2009;**14**:680-688

[23] Malhova E. Cytoembryology du genre *Capsicum*. Eucarpia Capsicum. 1977;**77**:191-197

[24] Indira C, Susan A. Morphological and cytological studies on a radiation induced polyploid in *Capsicum annum* Linn. Cytologia. 1977;**42**:371-375

[25] Raghuvanshi SS, Sheila J. Cytomorphological studies on the colchiploids of *Capsicum frutescens* L. Cytologia. 1964;**29**:61-78

[26] Takizawa K, Ishikawa K, Nunomura O, Ito T. Ploidy level effect on physiology of pepper plant as affected by fruit loading. Acta Horticulturae. 2008;**779**:689-697

[27] Watts L. Polyploidy. In: Watts L, editor. Flower and Vegetable Plant Breeding. London: Grower Books; 1980. pp. 22

[28] Ishikawa K, Mishiba K, Yoshida H, Nunomura O. Establishment of tetraploid plants of *Capsicum annuum* L. by colchicine treatment with the analysis of flow cytometry. Capsicum and Eggplant Newsletter. 1997;**16**:44-47

[29] Ishikawa K. Tetraploid bell pepper shows high in vitro pollen germination at 15°C. HortScience. 2001;**36**(7):1336

[30] Pal BP, Ramanujan S, Joshi AB. Colchicine induced polyploidy in crop-plants. II Chilli (*Capsicum annuum* L.). Indian Journal of Genetics and Plant Breeding. 1941;**1**:28-40

[31] Rao KG. Colchicine induced chromosome mosaicism in chili pepper (*Capsicum annuum* L.). Proceedings of the Indian Academy of Sciences. 1987;**97**(1):55-61

[32] Hendy H, Pochard E, Dalmasso A. Transmission héréditaire de la résistance aux Meloidogyne portée par deux lignées de *Capsicum annuum*: études de descendances d'homozygotes issues d'androgénèse. Agronomie. 1985;**5**:93-100

[33] Palloix A. Diseases of pepper and perspectives for genetic control. In: Capsicum Newsletter Special Issue-Proc VIIIth EUCARPIA Meeting on Genetics and Breeding of Capsicum and Eggplant. Rome, Italy; 1992. pp. 120-126

[34] Pochard E. Description of trisomic individuals of *Capsicum annuum* L. obtained in progeny of a haploid plant. Ann Amelior Plantes. 1970;**20**:233-256

[35] Pochard E. Localisation of genes in *Capsicum annuum* L. by trisomic analysis. Ann Amelior Plantes. 1977;**27**:255-266

[36] Tanksley SD. Linkage relationships and chromosomal locations of enzyme-coding genes in pepper (*Capsicum annuum* L.). Chromosoma. 1984;**89**:353-360

[37] Pickersgill B. Chromosomes and evolution in *Capsicum*. In: Pochard E, editor. Capsicum 77. Comptes Rendues 3'eme Congre's Eucarpia Piment. Avignon, France: Montfavet; 1977. pp. 27-37

[38] Fari MG, Csillery, Zatyko L. Embryo culture: An efficient technique in interspecific hybridization and in breeding of pepper (*Capsicum*). In: EUCARPIA Meeting on Genetics and Breeding of Capsicum and Eggplant. Plovdiv, Bulgaria; 1983. pp. 31-37

[39] Shivanna KR, Bahadur B. Efficacy of biotechnological approaches to raise wide sexual hybrids. In: Bahadur B, Rajam MV, Sahijram L, Krishnamurthy KV, editors. Plant Biology and Biotechnology. New Delhi: Springer India; 2015. pp. 347-362

[40] Yoon JB, Yang DC, Do JW, Park HG. Overcoming two post-fertilization genetic barriers in interspecific hybridization between *Capsicum annuum* and *C. baccatum* for introgression of anthracnose resistance. Breeding Science. 2006;**56**:31-38

[41] Manzur JB, Fita A, Prohens J, Rodríguez-Burruezo A. Successful wide hybridization and introgression breeding in a diverse set of common peppers (*Capsicum annuum*) using different cultivated Ají (*C. baccatum*) accessions as donor parents. PLoS One. 2015;**10**(12):1-18

[42] Shen X, Gmitter FG Jr, Grosser JW. Immature embryo rescue and culture. In: Thorpe TA, Young EC, editors. Plant Embryo Culture. New York: Humana Press; 2011. pp. 75-92

[43] Pickersgill B. The genus *Capsicum*: A multidisciplinary approach to the taxonomy of cultivated and wild plants. Biol Zent. 1988;**107**:381-389

[44] George L, Narayanaswamy S. Haploid *Capsicum* through experimental androgenesis. Protoplasma. 1973;**78**:467-480

[45] Kuo JS, Wang YY, Chien NF, Ku SJ, Kung ML, Hsu HC. Investigations on the anther culture in vitro of *Nicotiana tabacum* and *Capsicum annuum*. Acta Botanica Sinica. 1973;**15**:36-52

[46] Alremi F, Taskın H, Sönmez K, Buyukalaca S, Ellialtioglu S. Effect of genotype and nutrient medium on anther culture of pepper (*Capsicum annuum* L.). Turkish Journal of Agricultural and Natural Sciences. 2014;**1**:108-116

[47] Grozeva S, Todorova V, Cholakov T, Rodeva V. Effect of temperature and growth period of donor plants on pepper anther culture. In: 3rd International Conference on Research People and Actual Tasks on Multidisciplinary Sciences. Vol. 1. Lozenec, Bulgaria; 2013. pp. 60-64

[48] Trajkova F, Koleva-Gudeva L. Fruit analysis of pepper androgenic lines P3 and P4 (*Capsicum annuum* L. cv. Piran) in different maturation stages. Yearbook 2014, Goce Delcev University—Stip, Faculty of Agriculture; 2014. pp. 51-66

[49] Chunling L, Baojun Y. Successful development of new sweet (hot) pepper cultivars by anther culture. Acta Horticulturae. 1995;**402**:442-444

[50] Kisiała A, Olszewska D, Niklas-Nowak A, Nowaczyk P. Biometrical

characteristics of R2 generation of anther-derived pepper (*Capsicum* spp.) plants. Acta Agrobotanica. 2011;**64**:53-58

[51] Shrestha LS, Luitel BP, Kang WH. Agromorphological characterization of anther derived plants in sweet pepper (*Capsicum annuum* L. cv. Boogie). Horticulture, Environment, and Biotechnology. 2011;**52**:196-203

[52] Luitel B, Kang W. In vitro androgenic response of minipaprika (*Capsicum annuum* L.) genotypes in different culture media. Horticulture, Environment, and Biotechnology. 2013;**54**:162-171

[53] Luitel B, Kang W. Assessment of fruit quality variation in doubled haploids of minipaprika (*Capsicum annuum* L.). Horticulture, Environment, and Biotechnology. 2013;**54**:257-265

[54] Nowaczyk L, Banach-Szott M, Olszewska D, Nowaczyk P. Androgenic response of *Capsicum* interspecific hybrids and capsaicinoid characteristics of DH lines. Herba Polon. 2014;**60**:50-59

[55] Hwang JK, Paek KY, Cho DH. Breeding of resistant pepper lines (*Capsicum annuum* L.) to bacterial spot (*Xanthomonas campestris* pv. *vesicatoria*) through anther culture. Acta Horticulturae. 1998;**461**:301-310

[56] Nervo G, Azzimonti MT, Bonelli A, Tamietti G. Application of in vitro anther culture methods to a pepper breeding program for disease resistance. In: Proceedings of 51st Italian Society of Agricultural Genetics Annual Congress. Riva del Granda, Italy; 2007

[57] Mitykó J, Gémes JA. Improvement in the haploid technique routinely used for breeding sweet and spice peppers in Hungary. Acta Agronomica Hungarica. 2006;**54**:203-219

[58] Todorova V, Grozeva S, Rodeva V, Masheva S. Breeding evaluation of pepper lines obtained by in vitro anther culture. Genetika. 2013;**45**:601-610

[59] Grozeva S, Rodeva V, Todorova V, Pundeva R. Obtaining of pepper plants via anther culture. Genetics and Breeding. 2009;**38**:25-31

[60] Liu W, Parrott WA, Hildebrand DF, Collins GB, Williams EG. Agrobacterium induced gall formation in bell pepper (*Capsicum annuum* L.) and formation of shoot-like structures expressing introduced genes. Plant Cell Reports. 1990;**9**:360-364

[61] Lee YH, Kim HS, Kim JY, Jung M, Park YS, Lee JS, et al. A new selection method for pepper transformation: Callus-mediated shoot formation. Plant Cell Reports. 2004;**23**:50-58

[62] Shin R, Park JM, An JM, Park KH. Ectopic expression of TsiI in transgenic hot pepper plants enhances host resistance to viral, bacterial and oomycete pathogens. Molecular Plant-Microbe Interactions. 2002;**15**:983-989

[63] Shin R, Han JH, Lee GJ, Peak KH. The potential use of a viral coat protein gene as a transgene screening marker and multiple virus resistance of pepper plants coexpressing coat proteins of cucumber mosaic virus and tomato mosaic virus. Transgenic Research. 2002;**11**:215-219

[64] Voinnet O. RNA silencing as a plant immunity system against viruses. Trends in Genetics. 2001;**17**:449-459

[65] Harpster MH, Brummell DA, Dunsmuir P. Suppression of a ripening-related endo-1,4-b-glucanase in transgenic pepper fruit does not prevent depolymerization of cell wall

polysaccharides during ripening. Plant Molecular Biology. 2002;**50**:345-355

[66] Kim DH, Kang JG, Kim S, Kim BD. Identification of coxII and atp6 region as associated to CMS in *Capsicum annuum* by using RFLP and long and accurate PCR. Journal of Korean Society of Horticultural Science. 2001;**42**:121-127

[67] Kim S, Kim SR, Chung Sun AN, Hong YN, Lee KW. Constitutive expression of rice MADS box gene using seed explants in hot pepper (*Capsicum annuum* L.). Molecules and Cells. 2001;**12**(2):221-226

[68] Koeda S, Sato K, Saito H, Nagano AJ, Yasugi M, Kudoh H, et al. Mutation in the putative ketoacyl-ACP reductase CaKR1 induces loss of pungency in *Capsicum*. Theoretical and Applied Genetics. 2019;**132**(1):65-80

[69] Chee M, Lycett GW, Foan C. Development of a direct transformation method by GFP screening and in vitro whole plant regeneration of *Capsicum frutescens* L. Electronic Journal of Biotechnology. 2018;**34**:51-58

[70] Collard BC, Mackill DJ. Marker-assisted selection: An approach for precision plant breeding in the twenty-first century. Philosophical Transactions of the Royal Society B. 2008;**363**:557-572

[71] Kole C, Gupta PK. Genome mapping and map based cloning. In: Jain HK, Kharkwal MC, editors. Plant Breeding: Mendelian to Molecular Approaches. New Delhi, India: Narosa Publishing House; 2004. p. 811

[72] Costa MP, Rêgo MM, da Silva AP, do Rêgo ER, Barroso PA. Characterization and genetic diversity of pepper (*Capsicum* spp) parents and interspecific hybrids. Genetics and Molecular Research. 2016;**15**:2

[73] Suwor P, Sanitchon J, Thummabenjapone P, Kumar S, Techawongstien S. Inheritance analysis of anthracnose resistance and marker-assisted selection in introgression populations of chilli (*Capsicum annuum* L.). Scientia Horticulturae. 2017;**220**:20-26

[74] Bhadragoudar MR, Patil CG. Assessment of genetic diversity among *Capsicum annuum* L. genotypes using RAPD markers. African Journal of Biotechnology. 2011;**10**(76):17477-17483

[75] Lafortune D, Béramis M, Daubèze AM, Boissot N, Palloix A. Partial 4 resistance of pepper to bacterial wilt is oligogenic and stable under tropical conditions. Plant Disease. 2005;**89**:501-506

[76] Ibiza VP, Blanca J, Canizares J, Nuez F. Taxonomy and genetic diversity of domesticated *Capsicum* species in the Andean region. Genetic Resources and Crop Evolution. 2012;**59**(6):1077-1088

[77] Taranto F, Agostino N, Greco B, Cardi T, Tripodi P. Genome-wide SNP discovery and population structure analysis in pepper (*C. annuum*) using genotyping by sequencing. BMC Genomics. 2016;**17**:943

[78] Cheng J, Zhao Z, Li B, Qin C, Wu Z, Trejo-Saavedra DL, et al. A comprehensive characterization of simple sequence repeats in pepper genomes provides valuable resources for marker development in *Capsicum*. Scientific Reports. 2016;**6**:18919

[79] Rao GU, Paran I. Polygalacturonase: A candidate gene for the soft flesh and deciduous fruit mutation in *Capsicum*. Plant Molecular Biology. 2003;**51**:135-141

[80] Ben Chaim A, Grube RC, Lapidot M, Jahn M, Paran I.

Identification of quantitative trait loci associated with resistance to cucumber mosaic virus in *Capsicum annuum*. Theoretical and Applied Genetics. 2001;**102**:1213-1220

[81] Chen C, Hao X, Chen G, Cao B, Chen Q, Liu S, et al. Characterization of a new male sterility-related gene CaMF1 in *Capsicum annuum* L. Molecular Biology Reports. 2012;**39**:737-744

[82] Tiwari A, Vivian-Smith A, Voorrips RE, Habets MEJ, Xue LB, Offringa R, et al. Parthenocarpic potential in *Capsicum annuum* L. is enhanced by carpelloid structures and controlled by a single recessive gene. BMC Plant Biology. 2011;**11**:143

[83] Kang WH, Huy NH, Yang HB, Kwon JK, Jo SH, Seo JK, et al. Molecular mapping and characterization of a single dominant gene controlling CMV resistance in peppers (*Capsicum annuum* L.). Theoretical and Applied Genetics. 2010;**120**:1587-1596

[84] Daskalov S. Mutation breeding in pepper. In: Micke et al, editors. Mutation Breeding Review. Vienna: International Atomic Energy Agency/FAO; 1986. p. 25

[85] Raghavan TS, Venkatasubban KR. Studies in the South Indian chillies. Proceedings: Plant Sciences. 1940;**12**:29-46

[86] Paran I, Borovsky Y, Nahon S, Cohen O. The use of induced mutations to study shoot architecture in *Capsicum*. Israel Journal of Plant Sciences. 2007;**55**:125-131

[87] Cohen O, Borovsky Y, David-Schwartz R, Paran I. *Capsicum annuum* S (CaS) promotes reproductive transition and is required for flower formation in pepper (*Capsicum annuum*). The New Phytologist. 2014;**202**:1014-1023

[88] Hwang D, Jeong HJ, Kwon JK, Kim H, Kang SY, Kang BC. Phenotypic variants among ethyl methane sulfonate M2 mutant lines in *Capsicum annuum*. Plant Genetic Resources. 2014;**12**:141-145

[89] Jeong HJ, Kwon JK, Pandeya D, Hwang J, Hoang NH, Bae JH, et al. A survey of natural and ethyl methane sulfonate-induced variations of eIF4E using high-resolution melting analysis in *Capsicum*. Molecular Breeding. 2012;**29**:349-360

[90] Daskalov S. A male sterile pepper (*C. annuum* L.) mutant. Theoretical and Applied Genetics. 1968;**38**:370-372

[91] Daskalov S. Investigation of induced mutants in *Capsicum annuum* L. III. Mutants in the variety Zlaten medal. Genetic Selection. 1973;**6**:419-429

[92] Honda I, Kikuchi K, Matsuno S, Fukuda M, Santo H, Ryuto N, et al. Effect of heavy ion bombardment on mutagenesis in sweet pepper isolated by M1 plant selection. Euphytica. 2006;**15**(1):61-66

[93] Daskalov S, Mihailov L. A new method for hybrid seed production based on cytoplasmic male sterility combined with a lethal gene and a female sterile pollenizer in *Capsicum annuum* L. Theoretical and Applied Genetics. 1988;**76**:530-532

Chapter 7

Postharvest Handling Methods, Processes and Practices for Pepper

Oluyinka Adewoyin, Amos Famaye, Rufus Ipinmoroti, Adebayo Ibidapo and Folasayo Fayose

Abstract

Pepper belongs to the family Solanaceae. It is cultivated for its pungency, flavor, color, taste, export potential, capsaicin and oleoresin content. It is classified as sweet or hot pepper, depending on the capsaicin content of the fruit. World pepper production was around 3.5 million tons between 2009 and 2019, with 35% from Vietnam as the largest producer and exporter, followed by India and Indonesia. Vietnam pepper production increased progressively by 12.4% in 2014. The country was the major producer, followed by Brazil, Indonesia and India. To enhance adequate supply of pepper fruits and stabilize its soaring price, there must be efficient postharvest handling, processing and preservation methods compatible with the socio-economic and cultural practices of the producer. Data showed that improper postharvest handling of pepper results in huge postharvest losses. With this, handling must be a crucial part of an integrated systematic approach to maintaining the final product's quality.

Keywords: pepper, postharvest handling methods, preservation, processing, practices

1. Introduction

Postharvest technology is an essential part of agricultural production and utilization system. It is vital in loss reduction, value addition, food security, employment and income generation. Therefore, there is an urgent need for a postharvest technology revolution with strong linkages to proper processing, preservation, storage, marketing and distribution of pepper fruits [1]. Inappropriate postharvest handlings of pepper result in huge losses [2]. Poor handling practices lead to substantial postharvest losses. These practices include;- harvesting at an improper maturity stage, wrong harvesting method, use of inappropriate field packaging materials, harvesting at an ill-chosen time of the day, poor transportation, inadequate storage system and improper processing and preservation methods. Ineffective postharvest handling will reduce fruits' shelf life, hasten postharvest decay and reduce marketability [3, 4]. Physiologically, pepper fruit lacks natural wax at an immature stage, which hastens moisture loss and reduces quality. Postharvest losses have been estimated to be about 26–35 per cent for local markets and 50 per cent for exports [5–8]. Successful storage requires a good product, proper temperature and atmospheric humidity, the right stage of maturity, proper harvesting method, right harvesting time, suitable sanitation procedure and freedom from diseases and injury. These factors enhance and determine the shelf life of pepper in storage, while impacts of damages such as bruises, rupture and puncture can be reduced by proper handling [9]. Pepper fruits are harvested, handled and stored just as it suits each individual. Postharvest handling method and storage of pepper fruits are based on demand and supply. The consequence is that the system has not led to necessary and efficient postharvest handling and storage of pepper fruits. Due to the absence of a proper postharvest management system, a bulk quantity of the fruits gets damaged during handling, transportation and marketing, resulting in substantial annual losses [10–13].

2. Pepper

Pepper is a herbaceous plant with a tap root system. It grows up to 10-20 mm long and 3-7 mm in diameter, and fruit weight ranges from 128 to 210 g depending upon cultivars. The fruit grows from green through pale yellow to mature bright red [14]. There are 26 different species of pepper, but the most widely cultivated are *Capsicum chinense*, *Capsicum frutescens*, *Capsicum annum*, *Capsicum baccatum* and *Capsicum pubescens* [15]. The centre of diversity for *Capsicum* species is in South-central and South America, with most species having the same range in Brazil and Bolivia [16]. The primary centre of origin for domesticated *C. annum* is in semi-tropical Mexico [17, 18]. Bolivia is considered the centre of domestication for *C. baccatum* (*Aji*) and *C. Pubescens* (*rocoto*). Pepper's diversity is associated with its different names; chilly, chily, mirchi, chili, chile, *Aji*, paprika and capsicum for plants in the genus capsicum [19]. Pepper can appear in the local dialects as follows:

Nigeria	
Yoruba	Rodo
Ibo	Osse nkrisi
Hausa	Barkono/Tasshi
Afrikaans	Brand rissie
Amharic	Mit'mita berbere
Arabic	Fulfill ahmar

2.1 Common names for capsicum

Aji: This is a South American term used for chili.

Anaheim: They are blunt-nosed, long and narrow with a green or reddish color.

Bolita: Bolita is dark red, oval-shaped chili fruit with very high pungency.

Cascabel: Dried form of bolita.

Cayenne chili: It is long and slender with dark green color. It turns red at maturity.

Chiltepin: It is one of the oldest and original chili species. It has tiny round-shaped fruit.

Habanero: They are well known in Mexico and USA; they have thick flesh and green color, cylindrical and oval-shaped, which turns red when ripe.

Jamaican Hot: It originated from Jamaica. The color is green when immature and turns to yellow color at maturity.

Mirasol: They are beautiful, smooth, shining red skin and oval-shaped with high pungency. When dried, it is known as *'guajillo'* or *'puya'*.

Paprika: Paprika belongs to Capsicum annum; it originated from Mexico. It is usually dried and ground for seasoning.

Pasilla chili: It is long and thin with green color, which turns dark brown when it matures; it is mild to medium in pungency. The ripe pasilla is called "*Chilaca*' or "Chiles Negro'.

Pimento: This is Spanish paprika called pimento chili or cherry pepper. It is extremely mild in pungency; it is large heart-shaped red pepper with sweet, aromatic and succulent fruit.

Poblano: It originated from Pueblo in Mexico. It is heart-shaped green in color when unripe and turns red or brown at maturity. The pungency ranges from mild to moderate. The dried poblanos are known as Mulato and Ancho.

Scot's Bonnet or Scotch Bonnet: They are very high in pungency and may cause blisters to the tongue, dizziness and severe heartburn. They are irregular in shape with yellow, orange or red color.

Serrano: They are small, round in shape and slightly pointed at the end. They are smooth dark green when unripe and turn scarlet red, brown, orange and then yellow as they ripen with high pungency. They originated from the foothills of Puebla in Mexico.

Birdseye or Dhani Chillies: It is also known as African Devil Chile. It is tiny, green and bright red at maturity. It is grown in many African countries.

Byadagi or Kaddi Chillies are grown in Goa and Dharwar in Karnataka. These chillies are also called Kaddi chillies and wrinkled red when dried.

Guntur Sanman Chillies: These chillies are cultivated in the Guntur, Warangal and Khammam district of Andra Pradesh. The Guntur chillies are long with thick red skin with very high pungency.

Hindpur Chillies: These are found in Andhra Pradesh in India. These chillies are extremely pungent and red. The harvesting season for these chillies is from December to March.

Jwala Chillies: The most popular form of chili in India. Jwala is long and slender; when unripe, it has green color and turns red when ripe. Jwala means "volcano' in Hindi and is highly pungent. Jwala is found in Kheda and Mehsana in Gujarat.

Kanthari Chillies: The Kanathari chillies are grown in Kerala and Tamil Nadu. They are small and ivory white in color. They are highly pungent.

Kashmiri Mirch: It is grown in Himachal Pradesh, Jammu and Kashmir. It has smooth, shining skin and is fleshy with dark red color. They are mild in pungency.

Mundu Chillies or Gundu Molzuka: Are found in Tamil Nadu and Anantpur in Andhra Pradesh. They are roundish fruit with a moderately pungent yellowish red color.

Nalcheti Chillies: They are grown in Nagpur and Maharashtra. They are long and red when dried with high pungency.

Tomato Chili or Warangal Chappatta: These are found in Warangal, Khammam and the Godavari district of Andra Pradesh. These chillies are short, dark red in color, and when dried, these chillies have moderate pungency.

2.1.1 Factors influencing the nutritional composition of pepper

2.1.1.1 Maturity stage and harvesting method

Pepper's maturity is in three distinct stages during their developmental stages. These stages are immature green, mature green and mature red. It starts from immature green to mature green, pepper fruit increases in firmness and pungency, the cell walls thicken and no color change occurs. Mature green peppers are horticulturally developed and can be consumed fresh or in processed form. Harvesting of immature green peppers results in poor color, flavors and short lifespan [20]. Maturity is a major factor determining the compositional quality of fruits and vegetables. Fruit maturity depends on the cultivar and the environmental conditions before and during maturation [21]. Various workers reported that immature citrus fruits contained the highest concentration of vitamin C, whereas ripe fruits contained the least. Although vitamin C concentration decreased during the maturation of citrus fruits, the total vitamin C content per fruit tended to increase because the total volume of juice and fruit size increased with advancing maturity [22]. The method of harvest, maturity and physical injuries can influence the nutritional composition of pepper fruits. Mechanical damages (e.g. bruising, surface abrasions and cuts) can result in accelerated loss of vitamin C. The incidence and severity of such injuries are influenced by the method of harvest and handling operations. Vitamin C loss occurs when vegetables are severely cut or shredded, such as cabbage, lettuce, carrots and other vegetables sold as salad mix. Proper management should be employed to minimize physical damage to crops, whether harvesting is done by hand or by machine. Strawberries and other berries lose vitamin C quickly if bruised during harvesting [23]. Mondy and Leja found a considerable decrease in vitamin C content of injured tissue of potato tubers, while the unbruised halves appeared to show an increase in their vitamin C content [24, 25]. The harvest and postharvest techniques adopted by small-scale farmers in some developing countries were inappropriate for protecting fruits from damage and deterioration. The farmers also sometimes fail to sort out infected fruits from wholesome ones before transporting them to market, which could have reduced infection of healthy fruits and subsequent postharvest losses. Other causes of high failures include using unskilled labour for harvest, careless loading and unloading of harvested fruits, exposure of fruits to direct sunlight resulting in heat buildup and poor roads and transportation networks. Adewoyin and Babatola revealed that the shelf-life of pepper fruits harvested with pedicels at 10% ripeness and stored in refrigerator, evaporative coolant structure (ECS) and ambient condition (AC) were 27, 20 and 6 days, while those harvested at 100% ripeness were 21, 14 and 3 days, respectively. The corresponding shelf-life of pepper harvested without pedicel at 10% ripeness was 21, 18 and 5 days and those harvested at 100% ripeness were 18, 15 and 2 days. Deterioration was significantly higher in fruits harvested without pedicels than those harvested with pedicels [26].

2.1.2 Climatic condition

During the growing season, the amount and intensity of light significantly influence the amount of vitamin C formed. Vitamin C is synthesized from sugars supplied through photosynthesis in plants. Outside fruit exposed to maximum sunlight contains more vitamin C than inside and shaded fruit on the same plant. It was observed that grapefruits grown in coastal California generally contain more vitamin C than fruit grown in desert areas.

2.1.3 Cultural practices

Plant growth is generally enhanced by Nitrogen fertilizer, so a relative dilution effect may occur in the plant tissues [23]. More so, cultural practices like pruning and thinning determine the crop load and fruit size, which can influence the nutritional composition of fruits. Using pesticides and growth regulators may indirectly affect the nutritional quality of fruit. The application of gibberellins was beneficial to green tea quality, increasing vitamin C content by 18%. Excess use of Nitrogen fertilizer increases the concentration of NO_3 and simultaneously decreases that of AA. It may have a double negative effect on the quality of plant food [27, 28]. Although vitamin C concentration is positively correlated with the nitrogen supply in butter-head lettuce, it is inversely correlated with the Nitrogen supply in white cabbage and crisp-head lettuce [29–32].

2.1.4 Temperature

Temperature is critical in extending shelf life and maintaining the quality of fresh pepper fruit. Prolonged keeping of fruits after harvesting and cooling or processing can result in direct loss due to water loss and decay; indirect losses can also occur, such as flavor and deterioration in nutritional quality. The temperature range and the extent of vitamin C loss depended on the type of fruit. Wu found that vitamin C reduced quickly in green beans kept at 5°C after three days but retains stability in broccoli [33, 34]. Esteve showed that vitamin C concentration in harvested fresh green asparagus stored at 4°C increased after two days [35].

2.1.5 Mechanical damage

Mechanical damages such as abrasion, cuts and bruises affect the chemical composition of pericarp tissues of tomato fruit. Vitamin C content was lower by 15% in bruised locular tissue than in fruit without physical damage. Shelling green peas and green lima beans lowers their nutrient composition compared to those left in the pod [36–38].

2.1.6 Chemical treatments

Calcium dips can control physiological disorders and firmness reduction in apples and cherries. Dehydrated pineapples and guava pre-treated with cysteine hydrochloride had increased AA retention and reduced color change during storage [39]. Kiwi fruit slices stored in ethylene–free air had threefold more AA content than the control [40].

2.1.7 Irradiation

Ionizing radiation can be used to prevent sprouting, insect control or delay of ripening of certain fruits and vegetables [41]. Irradiation effect on horticultural crops at relatively low doses at 300gy had no significant effect on AA and DHA. Irradiation at 75 – 100gy irreversibly inhibited the sprouting of potatoes regardless of storage temperature. Losses in vitamin C were lower in potato irradiated for sprout control and subsequently stored at 15°C than in non-irradiated tubers stored at 2-4°C [42]. 'Galia' musk melons were irradiated at doses up to 1kgy as a quarantine treatment, and the treatment did not affect vitamin C content. In general, doses of 2–3 kg combined with refrigeration helped extend the shelf life of strawberries [43]. During storage, AA levels significantly increased while DHA content decreased in irradiated strawberries [40].

2.1.8 Controlled modified atmosphere

Modifying the atmospheric air (AA) during storage reduces physiological and chemical changes in fruits. For instance, AA loss can be reduced by storing apples in a low oxygen atmosphere. Veltma found out that storage of pepper for six days in a

CO2-enriched atmosphere resulted in a reduction in AA content of sweet pepper kept at 13°C. Increasing CO_2 concentration in the storage atmosphere of strawberries had little effect on vitamin C content; AA was more diminished at high CO_2 than DHA [43]. Elevated CO_2 may stimulate the oxidation of AA by ascorbate peroxidase [44]. Other qualities of pepper were maintained better in MAP (Modified atmosphere package) than in air [43–49].

2.1.9 Processing methods

Processing methods are critical in preserving the beneficial properties of capsicum. Among the vitamins, ascorbic acid is very susceptible to chemical and enzymatic oxidation during the processing, cooking and storing of produce. Unblanched beans and pepper lost more than 97% of their vitamin C within one month of freezing at 23°C. Blanching reduces vitamin C content by 28% in vacuum-sealed samples further decreased by 3%, while non-vacuum sealed lost 10% in 12 months of storage [50–53]

2.2 Storage

The need for storage of agricultural produce is necessitated by seasonal changes, disasters, economic meltdown, climate change, price fluctuations, terrorism and tribal conflict. Tropical fruits are stored at a higher temperature than in the temperate region. The higher temperature increases metabolism, activities, respiratory rate, loss of moisture content and an increased rate of ripening; these reduce the shelf life of tropical fruits more rapidly. It was estimated that 25–70 per cent of the fresh fruit produced is lost after harvest [26]. Peppers stored above 7.5°C suffer water loss and shriveling. Storage below 7.5°C is best for a maximum shelf life of 3–5 weeks. Pepper dried to safe moisture content, packed tightly in sacks, can be stored in non-refrigerated warehouses for up to 6 months. Storage under low temperature reduces loss of red color and slow down insect activities. Rapid precooling of harvested pepper is essential in reducing marketing losses, and this can be done by forced air cooling, hydro cooling or vacuum cooling. If hydro cooling is used, care should be taken to prevent the development of decay. The moisture content of pepper should be kept low (10–15%) to prevent mold growth. Relative humidity of 60–70% is too higher. If the relative humidity is too low (below 10%), pods may be too brittle that they may shatter during handling, resulting in losses and the release of dust, which irritates the skin and respiratory system [54]. Using polyethene bags for dried pepper allowed for better storage and reduced dust. Packaging ensures the pod maintains a constant moisture content during storage until the grinding time. Baryeh studied the storage condition and storage life of okra and pepper in isolated concrete and heat storage unit in form of tubes and slabs. Solar heated air and the air blown around a hot internal combustion engine were passed through the tubes and slabs. The effectiveness of heat storage increased with increasing food storage length. It was demonstrated that up to 6.0 kj and 5.75 kj of heat energy could be used to store products in tropical semi-argillous mud and concrete over a 5-hr period. In a further study, concrete and mud tubes were used to store pepper, which was kept for one year without severe deterioration. The moisture content remains at the initial value of 10%, which was too low to allow germination and fungal growth [55]. Pepper can also be preserved effectively in the evaporative coolant structure for two weeks without problems associated with cold storage [56]. The objective of storage after harvest is to control the rate of deterioration and ripening, and to maintain its quality. The following characteristics are used as indices to measure the quality of pepper fruit; size, color, shape, firmness, freedom from defects, cracks, decay, sunburn, infection, physiological disorder and pathological disorder. Relative humidity, the control of relative humidity in a postharvest environment, is often as important as temperature control. In some situations, the effect of the two factors is difficult to separate because of the capacity of air to hold moisture, which varies with temperature. Relative humidity depends on the type of storage and purpose [54]. Green pepper stores best at 5–10°C for 3–4 weeks [57]. Wills also found that, in general, there is an inverse relationship between respiration rate and storage life so that produce with low respiration rate generally keep longer. It suggests that respiration could be slowed by limiting the oxygen or raising carbon dioxide concentration in the storage atmosphere. The reduction in the concentration of O_2 is necessary to achieve a dependent on the storage temperature. As the temperature lowers, the required concentration of O_2 is also reduced. For good quality, colored peppers should have 50% colouration [4].

2.3 Preservation techniques

The preservation of pepper helps in the following ways: To increase the economic value of the product by transforming it from one form to another, to remove inedible parts of produce, to improve taste, to inactivate enzymes, to make packaging easier, to reduce the cost of transportation, to make food available where it is not produced, to increase profit and to Control shortage.

2.3.1 Drying of foods

Drying is a food preservation method that involves the removal of moisture from food to a safe moisture level that inactivates microorganisms. Sun drying is only possible when the sun is available for a longer time [52]. Freeze-drying involves sublimation by converting food into ice without allowing water but through vacuum and heat applied in the drying chamber. The produce is frozen, and then water is removed by vacuum and application of heat. This coincides in the same chamber binding the moisture in food. High sugar concentration can also bind up the moisture and give the food a certain level of humidity at which micro-organisms cannot grow. Salt with high concentration can be used, resulting in high osmotic pressure that ties up the moisture and consequently inhibiting the growth of micro-organisms. It dehydrates the food by drying out and tying up moisture as it drains the microorganism's cells [58]. Pepper fruits parboiled at 75°C for 3 minutes before sun-drying had the highest nutrient composition with respect to fat, protein and capsaicin contents among various treatments evaluated. The efficient conservation of nutrients could be attributed to the inactivation of the enzyme system and retardation of physiological processes by parboiling which consequently prevents deterioration and nutrient losses. Parboiling has been found to reduce the period of drying and the maintenance of the red color of fresh pepper [59]. During drying, two processes occur – heat application and moisture evaporation from the sample. Nutrient losses were due to a more application of heat than the removal of moisture, which increased the concentration of nutrients in the fruit.

Using oven and dry parboiled, pepper fruits accelerated the moisture removal but resulted in lower nutritional composition than sun drying. Pepper fruits subjected to heat sterilization had high crude fibre, ash content and moisture content and had the lowest crude protein, fat, carbohydrate, capsaicin and oleoresin content. The ash content was highest in sun-dried pepper fruit. This agreed with the findings of

Mepba, who observed that the ash content of sun-dried vegetables was higher than the blanched samples. The parboiled pepper fruits had higher oleoresin and capsaicin content [52]. The crude fiber content of pepper fruits parboiled before oven drying or sun drying was significantly higher than those of the respective ones dried without parboiling. Oleoresin was positively correlated with all proximate content of pepper fruits except crude protein and fat content.

Furthermore, carbohydrate was correlated with protein, fat and fiber, while crude fiber was also linked to ash content. Fat content was also associated with moisture content and crude protein. In this study, mineral element content varied among treatments; Pepper parboiled before sun-drying or oven-drying had maximum values for all the mineral elements viz.: phosphorus, zinc, iron, potassium, calcium and sodium content. The mineral elements and vitamin C composition of pepper parboiled before sun-drying were significantly higher than those of the corresponding treatments. Heat-sterilized samples had the lowest phosphorus, calcium and vitamin C content than those subjected to various drying methods. Solanke and Awonorin (2002) have also reported losses of 62 to 93% of vitamin C in cooked vegetables [60]. The high solubility of vitamin C in water and the relative ease with which it is oxidized renders it susceptible to deterioration during processing. The route and rate of oxidation of vitamin C are influenced by several factors such as pH, presence of trace metals, enzymes, oxygen availability, time and temperature. The relationship of each nutrient element evaluated indicated a strong correlation among the treatments. Phosphorus content was highly correlated with calcium, zinc, iron, potassium, and sodium. Calcium was highly correlated with zinc, iron, potassium and sodium, while magnesium was moderately correlated with zinc and iron. Vitamin C was highly associated with potassium. The relationship between nutrients and vitamin C was all positive. Zinc was also highly correlated with iron, potassium and sodium, while potassium was moderately related to sodium.

2.4 Packaging and sales

The ultimate goal of packaging is to keep pepper fruits in good condition until they are sold to consumers (**Figures 1–3**). Pepper fruits are packaged to protect them from injury and water loss and be convenient for handling and marketing [27]. Packages should also provide information about the produce, including the grade, handling instructions and appropriate storage temperatures. Improper packaging and transportation methods, lack of shade or precooling facilities to remove field heat to reduce respiration rate and consequent deterioration rate are the reality among small-scale farmers in most developing nations. In Nigeria, for example, Adewoyin and Babatola observed that none of the marketers and farmers used the plastic crate designed by Food and Agriculture Organization (FAO) and Nigerian Stored Product Research Institute (NSPRI) to prevent physical damage to produce. This corroborates the findings of Olayemi *et al.* (2010) on the assessment of postharvest challenges of small-scale farm holders of tomatoes, bell and hot pepper in some local government areas of Kano State, who observed that peppers were typically harvested at a fully ripe stage (90%) and that most farmers still use the traditional basket and sack as their packaging material in conveying produce, which resulted in massive post-harvest losses of about 62.5%. Harvesting cardboard has been observed to be much more appropriate than polypropylene bags for fresh fruits. Harvesting cardboard boxes can go a long way in maintaining the quality of peppers for the following reasons: They bear the weight of the product compared to second-hand grain sacks, where the peppers carry the weight causing severe damage. They are easier to handle and can maintain the integrity of the fruits. Bruising, compression and friction are reduced. **Figure 1** showed the marketers' pepper using different consumer packages. The most common packaging materials used by the handlers were baskets of various sizes. Its limitations include injury and bruises due to hard rough surfaces during

Figure 1.
Display of pepper fruits by handlers [Adewoyin O.B].

Figure 2.
Improper packaging of pepper fruits [Adewoyin O.B].

Figure 3.
Packaged peppers ready for transport [Adewoyin O.B].

transportation, loading and handling after the close of the day. A survey by Adewoyin and Babatola on postharvest handling of pepper fruits by marketers showed that Hausa men were more involved in marketing pepper fruits than women. This could be attributed to the rigorous demand of the business that required constant transportation of pepper fruits from the north to the southwest. The religious background and customs in the north also restricted women to more domestic activities within the home environment (**Figures 1** and **2**) [61].

Figures 2 and **3** showed the bulk handling of pepper using a polypropylene bag, which has the following limitation:

1. Peppers are tightly packed in the bag, leading to compressions, lack of proper ventilation and heat buildup. This practice results in physical and mechanical damage and enzyme reactions along the marketing chain.

2. The temperature in the bags gets very high, which leads to rapid moisture loss and shriveling.

3. The bags are placed on top of each other during transportation resulting in further fruit damage.

4. Polypropylene bags may compromise food safety because they are not sanitized after use and are not stored under conditions that ensure food safety.

5. The netted nature of the bags can severely bruise the fruits during direct.

2.5 Uses and application

Capsicum is used for food flavoring, as a coloring agent, and as an essential condiment in food. Vitamins A, C and E are also present in its fruit. The green fruit of capsicum contains three times the vitamin C content of an orange and provides the minimum daily requirement for humans [16]. As the green pod turns red, pro-vitamin A content increases until it contains twice the pro-vitamin A of a carrot. Half a tablespoon of pepper powder furnishes the minimum daily requirement of vitamin A [14]. *Capsicum* coating and powder are used to control browsing animals and insects. The acetone and petroleum ether extracts caused complete mortality of rice weevil (*Sitophillus oryzae*) in 15 days, while fruit powder was much less effective. As a pharmaceutical ingredient, its usage has grown exceedingly. It has been used in the neurogenic bladder, osteo-arthritis-psoriasis, diabetes and neuropathy [20]. It is used for cluster headaches, indigestion, heartburn, itching, obesity, rheumatoid, arthritis, bursitis, migraine and headache. It is also utilized as a counter-irritant for asthma, coughs, sore throat and relieving toothache. It relieves itching in dialysis patients; reduces long-time inflammation [21]. Red or hot peppers from *Capsicum annum* and *Capsicum frutescens* are used extensively in Mexican and Italian food. As a traditional medicinal plant, capsicum has been used as a calmative, digestive irritant, stomach stimulant, rubefacient and tonic. Other purposes of capsicum include folk remedies for dropsy, colic, diarrhea, muscle cramp and toothache. *C. frutescens* L has been reported to have hypoglycemic properties and can cause contact dermatitis and blisters. Excessive consumption can cause gastroenteritis and kidney damage. Pepper may aggravate symptoms of duodenal ulcers. With increased body temperature, the flow of saliva and gastric juices may be stimulated by capsicum pepper. The medicinal application of capsaicinoids has brought innovative ideas for their use. Medicinal use of capsicum has a long history dating back to the Mayas, who used it to treat cough and sore throats. The Aztecs used chile pungency to relieve toothaches. The pharmaceutical industry uses capsaicin as a counter-irritant balm for external applications [21]. Pepper is the active ingredient in heat and Sloan's liniment to rub down liniments, used for sore muscle. It is used as an antiseptic, topical vasodilator, neural stimulant and depressant; diaphoretic for the first stage of cold when the skin is hot and dry has a protective effect on the tracheobronchial system. Capsaicin was determined to inhibit the growth of the gastric pathogen *Helicobacter pylori,* which is associated with gastric ulcers [22].

3. Conclusion

Postharvest technology has become necessary to improve food safety and strengthen the nation's food security. Appropriate handling, packaging, transportation and storage reduce post-harvest losses of pepper. For every one per cent reduction in loss, it will correspondingly save five million tons of pepper fruit annually. Processing and preservation technology helps to save excess fruit during the glut season. The technology helps to boost the export of pepper in the form of preserved and value-added products. The technical activities that must be incorporated in the pepper's value chain include harvesting at the proper maturity stage, removal of field heat, appropriate packaging, ventilated and sound vehicles, field drying, storage, proper processing, marketing and distribution. The economic activities in the value chain of pepper, such as transporting, marketing, quality control, nutrition and extension services, serve as a means of employing a significant number of people, improving household income and increasing the nation's economy.

Author details

Oluyinka Adewoyin[1*], Amos Famaye[2], Rufus Ipinmoroti[2], Adebayo Ibidapo[3] and Folasayo Fayose[4]

1 Department of Crop Science and Horticulture, Federal University Oye, Oye-Ekiti, Nigeria

2 Cocoa Research Institute of Nigeria, Ibadan, Nigeria

3 Department of Hospitality and Tourism Management, Federal University Oye, Oye-Ekiti, Nigeria

4 Department of Engineering and Bio - Resources, Federal University Oye, Oye-Ekiti, Nigeria

*Address all correspondence to: oluyinka.adewoyin@fuoye.edu.ng

References

[1] Odeyemi OO. 2013:Post-harvest food problems in Nigeria: Kinds and Magnitude of Spoilage and Agents of Deterioration. In: A 3-day workshop on reduction of Postharvest food losses in Nigeria. Nigeria: FUTA; 2013

[2] Kader AA. Increasing food availability by reducing post-harvest losses of produce. Acta Horticulturae (ISHS). 2005;**682**:2169-2176

[3] Kitinoja L, Cantwell M. Identification of Appropriate Post-Harvest Technologies for Improving Market Access and Incomes for Small Horticultural Farmers in Sub-Saharan Africa and South Asia. WFLO Grant Final Report to the Bill and Melinda Gates Foundation. World Food Logistics Organization; 2010

[4] Brecht J, Bergsma K, Sanchez C, Snyder G. Harvest maturity and storage temperature effects on quality of Chinese, water chestnut. Acta Horticulture. 1992;**318**:313-319

[5] Kughur PG, Iornenge GM, Ityonongu BE. Effects of post-harvest losses on selected fruits and vegetables among small-scale farmers in Gboko Local Government area of Benue State, Nigeria. International Journal of Innovation and Scientific Research. 2015;**19**(1):201-208

[6] Kumrul H, Bishan L, Akhter N. Post-Harvest Loss Assessment: A Study to Formulate Policy for Loss Reduction of Fruits and Vegetables and Socioeconomic Uplift of the Stakeholders. National Food Policy Capacity Strengthening Programme Final Report. 2010. p. 188

[7] Verma A, Singh KP. An economic analysis of post-harvest losses in fresh vegetables. Indian Journal of Agricultural Marketing. 2004;**18**(1):136-139

[8] Vishal S, Hedayetullah MD, Parveen Z, Jagamohan M. Post-harvest technology of fruits and vegetables: An overview. Journal of Post-harvest Technology. 2014;**2**(2):124-135

[9] Adugna D, Gerba D, Diriba B, Kassaye T. Identification of major causes of postharvest losses among selected fruits in Jimma for proffering veritable solutions. International Journal of Current Research. 2011;**3**:34-40

[10] Ngowi ER, Selejio O. Post-harvest loss and adoption of improved post-harvest storage technologies by smallholder maize farmers in Tanzania. African Journal of Economic Review. 2019;**7**(1):249-267

[11] Kodwo IGA, Bright KDT. Determinants of Yam Post-Harvest Management in the Zabzugu District of Northern Ghana. Hindawi Publishing Corporation; 2016. p. 9

[12] Kumar KK, Basavaraja H, Mahajanshetti SB. An economic analysis of post-harvest losses of vegetables in Karnataka. Indian Journal of Agricultural Economy. 2006;**61**(1):134-146

[13] Hodges RJ, Buzby JC, Bennett B. Post-harvest losses and waste in developed and less developed countries: Opportunities to improve resource use. The Journal of Agricultural Science. 2011;**149**:37-45

[14] Bosland PW, Votava EJ. Peppers: Vegetable and Spice Capsicums. Crop Production Science in Horticulture. Wallingford: CAB International Publishing; 2000. p. 204

[15] Eshbaugh WH. Peppers: History and exploitation of a serendipitous new crop

discovery. In: Janick J, Simon JE, editors. New Crops. New York: John Wiley& Sons; 1993. pp. 132-139

[16] Bosland PW. Chiles; history, cutivation, and uses. In: Charalambous G, editor. Spicies, Herbs and Edible Fungi, Developments in Food Science. Vol. 34. Amsterdam: Elsevier; 1994. pp. 347-366

[17] Hernandez-Verdugo S, Luna-Reyes R, Oyama K. Genetic structure and differentiation of wild and domesticated population of Capsicum annuum (Solanaceae) from Mexico. Plant Systematics and Evolution. 2001;**226**:129-142

[18] Whitmore TM, Turner BL. Cultivated Landscapes of Middle America on the Eve of Conquest. New York: Oxford University Press; 2002. p. 338

[19] Bosland PW. Capsicums: Innovative uses of an ancient crop. In: Janick J, editor. Progress in New Crops. Arlington, Virginia, USA: ASHS Press; 1996. pp. 479-487

[20] Lynn B. Capsaicin actions on nociceptive C- fibers and therapeutic potential. Pain. 1990;**41**:61-69

[21] Chamichael JK. Treatment of herpes zoster and post therpetic neuralgia. Am. Family physician, pp. 1991;**44**:203-210

[22] Sharol Tilgner. Cayenne (Capsicumfrutescens). 1999. http://www.herbaltransitions.com/materiamedica/Capsicum.htm. [Accessed: June 7, 2013]

[23] Mondy NI, Leja M. Effect of mechanical injury on the ascorbic acid content of potatoes. Journal of Food Science. 1986;**51**:355-357

[24] Lasrizal MA, Giraud DW, Driskcll JA. Retention of vitamin C, iron, and beta-carotene in vegetables prepared using different cooking methods. Journal of Food Quality. 1997;**20**:403-418

[25] Howard LR, Smith RT, Wagner AB, Villalon B, Burns EE. Provitamin A and ascorbic acid content of fresh pepper cultivars (Capsicum animum) and processed jalapenos. Journal of Food Science. 1994;**59**:362-365

[26] Adewoyin OB. Effect of Postharvest Handling Methods and Storage Conditions on Storability of Pepper Fruit. Nigeria: University of Ibadan; 2014

[27] Lisiewska Z, Kmiecik W. Effect of level of nitrogen fertilizer, processing conditions and period of storage for frozen broccoli and cauliflower on vitamin C retention. Food Chemistry. 1996;**57**:267-270

[28] Mozafar A. Nitrogen fertilizers and the amount of vitamins in plants: A review. Journal of Plant Nutrition. 1993;**16**:2479-2506

[29] Freeman S, Toivonen PM, Perrin PW, Tin WC, Hall JW. Effect of nitrogen fertilization on yield, storage losses and chemical composition of winter cabbage. Canadian Journal of Plant Science. 1991;**71**:943-946

[30] Sorensen JN, Johansen AS, Poulsen N. Influence of growth conditions on the value of crisphead lettuce. Marketable and nutritional quality as affected by nitrogen supply, cultivar and plant age. Plant Foods for Human Nutrition. 1994;**46**:1-11

[31] Liang Y, Lu J, Shang S. Effect of gibberellins on chemical composition and quality of tea (Camellia sinensis L.). Journal of the Science of Food and Agriculture. 1996;**72**:411-414

[32] Albrecht JA, Schafer HW, Zottola EA. Sulphydryl and ascorbic

acid relationships in selected vegetables and fruits. Journal of Food Science. 1990;**56**(427):430

[33] Wu JQ, Schwarlz SJ, Carroll DE. Chemical physical and sensory stabilities of prebaked frozen sweet potatoes. Journal of Food Science. 1991;**56**:710-713

[34] Estrada B, Bernal MA, Diaz J, Pomar T, Merino F. Capsaicinoids in vegetative organs of Capsicum annum L. in relation to fruiting. Journal of Agricultural and Food Chemistry. 2002;50:1188-1191

[35] Nunes MCN, Brecht JK, Morais AMMB, Sargent SA. Controlling temperature and water loss to maintain ascorbic acid in strawberries during postharvest handling. Journal of Food Science. 1998;**63**:1033-1036

[36] Moretti CL, Sargent SA, Huber D, Calbo AG, Puschmann R. Chemical composition and physical properties of pericarp, locule, and placental tissues of tomatoes with internal bruising. Journal of the American Society for Horticultural Science. 1998;**123**:656-660

[37] Barry-Ryan C, O'Beirne D. Ascorbic acid retention in shredded iceberg lettuce as affected by minimal processing. Journal of Food Science. 1999;**64**:498-500

[38] Klieber A, Franklin B. Ascorbic acid content of minimally processed Chinese cabbage. Acta Horticulturae. 2000;**518**:201-204

[39] Mohammed S, Kyi KMM, Sharif ZM. Protective effect of cysteine – on vitamin C in dehydrated pickled/candied pineapples and guava. Journal of the Science of Food and Agriculture. 1993;**61**:133-136

[40] Agar IT, Streif J, Bangerth F. Effect of high CO_2 and controlled atmosphere on the ascorbic and dehy-droaseorbic acid content of some berry fruits. Postharvest Biology and Technology. 1997;**11**:47-55

[41] Mitchel GE, McLauchlan RL, Isaacs AR, Williams DJ, Nottingham SM. Effect of low dose irradiation on composition of tropical fruits and vegetables. Journal of Food Compound Analyses. 1992;**5**:291-311

[42] Joshi MR, Srirangarajan AN, Thomas P. Effects of gamma irradiation and temperature on sugar and vitamin C changes in five Indian potato cultivsars during storage. Food Chemistry. 1990;**35**:209-216

[43] Graham WD, Stevenson MH. Effect of irradiation on vitamin C content of strawberries and potatoes in combination with storage and with further cooking in potatoes. Journal of the Science of Food and Agriculture. 1997;**75**:371-377

[44] Wright KP, Kader AA. Effect of slicing and controlled-atmosphere storage on the ascorbate content and quality of strawberries and persimmons. Postharvest Biology and Technology. 1997;**10**:39-48

[45] Veltman RH, Sanders MG, Persijn ST, Peppelenbos HW, Oostcrhaven J. Decreased ascorbic acid levels and brown care development in pears (*Pyrus communis* L.). Physiology of Plant. 1999;**107**:39-45

[46] Paradis C, Castaigne F, Desrosiers T, Fortin J, Rodrigue N, Willemot C. Sensory, nutrient and chlorophyll changes in broccoli florets during controlled atmosphere storage. Journal of Food Quality. 1996;**19**:303-316

[47] Howard LR, Hernandez-Brenes C. Antioxidant content and market quality of jalapeno pepper rings as affected by minimal processing and modified

atmosphere packaging. Journal of Food Quality. 1998;**21**:317-327

[48] Gil MI, Ferreres F, Tomas-Barberan FA. Effect of modified atmosphere packaging on the flavonoids and vitamin C content of minimally processed Swiss chard (Beta vulgaris subspecies cycla). Journal of Agriculture Food Chemistry. 1998;**46**:2007-2012

[49] Miller N, Rice-Evans CA. The relative contributions of ascorbic acid and phenolic antioxidants to the total antioxidant activity of orange and apple fruit juices and blackcurrant drink. Food Chemistry. 1997;**60**:331-337

[50] Hagg M, Hakkinen U, Kumpulainen J, Ahvenainen R, Hurme E. Effects of preparation procedures, packaging and storage on nutrient retention in peeled potatoes. Journal of Science Food Agriculture. 1998;**77**:519-526

[51] Oruna-Concha MJ, Gonzalez-Castro MJ, Lopez-Hernan-dez J, Simal-Lozano J. Monitoring of the vitamin C content of frozen green beans and Padron peppers by HPEC. Journal of the Science of Food and Agriculture. 1998;**76**:477-480

[52] Mepba HD, Eboh L, Banigo DEB. Effects of processing treatments on the nutritive composition and consumer acceptance of some Nigerian edible leaf vegetables. African Journal of Food Agriculture Nutrition and Development. 2007;**7**:1

[53] Daramola AM, Okoye WI. Postharvest losses prevention in horticultural crop produce. In: Proceedings of 16th annual conference of Horticultural Society of Nigeria. Abeokuta, Nigeria: Societal Journal for Horticultural Society of Nigeria; 1998. pp. 12-16

[54] Willis RD, Mc Glasson H, Graham ND, Hall EG. Postharvest: An Introduction to the Physiology and Handling of Fruit and Vegetable B. Oxford, London Edinburgh: Sp Professional; 1989

[55] Baryeh EA. Thermal energy storage in concrete and mud for crop storage. Agricultural-mechanization in Asia, Africa and Latin America, Japan. 1990;**1990**:48-54

[56] Barbosa GV, Juan JF, Alzamora SM, Tapia MS, Aurelio L, Welti J. Handling and preservation of fruits and vegetables by combined methods for rural areas: A technical manual. FAO Agricultural Services Bulletin. 2003;**149**:99

[57] Davey MW, Van Montagu M, et al. Plant L-ascorbic acid: Chemistry, function, metabolism, bioavailability and effects of processing. Journal of the Science of Food and Agriculture. 2000;**80**:825-860

[58] Banaras M, Bosland PW, Lownds NK. Effects of harvest time and growth conditions on storage and post-storage quality of fresh peppers (Capsicum annuum L.). Pakistan Journal of Botany. 2005;**37**(2):337-344

[59] Dandamrongrak R, Mason R, Young G. The effect of pretreatments on the drying rate and quality of dried bananas. International Journal of Food Science and Technology. 2003;**38**:877-882

[60] Solanke OF, Awonorin SO. Kinetics of vitamin C degradation in some tropical green leaf vegetables during blanching. Nigerian Food Journal. 2002;**20**:24-32

[61] Montes HS, Garcia H, Gomez JAA. Fenologia del cultivo del chile (*Capsicum annuum* L.). In: Primera Convencion Mundial del Chile. Guanajuato, Mexico: Leony Celaya; 2004. pp. 43-48